Reasoning in Measurement

T0251889

This collection offers a new understanding of the epistemology of measurement. The interdisciplinary volume explores how measurements are produced, for example, in astronomy and seismology, in studies of human sexuality and ecology, in brain imaging and intelligence testing. It considers photography as a measurement technology and Henry David Thoreau's poetic measures as closing the gap between mind and world.

By focusing on measurements as the hard-won results of conceptual as well as technical operations, the authors of the book no longer presuppose that measurement is always and exclusively a means of representing some feature of a target object or entity. Measurement also provides knowledge about the degree to which things have been standardized or harmonized—it is an indicator of how closely human practices are attuned to each other and the world.

Nicola Mößner is a Junior Fellow at Alfried Krupp Wissenschaftskolleg Greifswald in Germany. She received her MA in German Literature and Linguistics at the University of Hamburg and her PhD in Philosophy at the University of Münster. Her thesis is about the epistemology of testimony and the special case of media reports, published as *Wissen aus dem Zeugnis anderer – der Sonderfall medialer Berichterstattung* (Paderborn: mentis 2010). She edited (together with Dimitri Liebsch) *Visualisierung und Erkenntnis – Bildverstehen und Bildverwenden in Natur- und Geisteswissenschaften* (Cologne: Herbert von Halem, 2012). Currently she is writing a book about conceptual and epistemological questions concerning visual representations in science. Her main research interests comprise philosophy of science and social epistemology.

Alfred Nordmann is Professor of Philosophy at the Technische Universität Darmstadt and adjunct professor at the University of South Carolina, Columbia. His interests in the philosophy of science concern the formation and contestation of fields of inquiry such as chemistry and theories of electricity in the 18th century, mechanics, evolutionary biology, and sociology in the 19th century. In particular, he has sought to articulate implicit concepts of science and objectivity. In 2000, he embarked on a similar endeavor in regard to nanoscience and converging technologies, which has led him to promote and develop a comprehensive philosophy of technoscience. Since the technosciences require new answers to the familiar questions of knowledge and objectivity, theory and evidence, explanation and validation, representation and experimentation, he is seeking to address these and related questions in his current work.

History and Philosophy of Technoscience
Series Editor: Alfred Nordmann

Titles in this series

1 **Error and Uncertainty in Scientific Practice**
 Marcel Boumans, Giora Hon and Arthur C. Petersen (eds)

2 **Experiments in Practice**
 Astrid Schwarz

3 **Philosophy, Computing and Information Science**
 Ruth Hagengruber and Uwe Riss (eds)

4 **Spaceship Earth in the Environmental Age, 1960 – 1990**
 Sabine Höhler

5 **The Future of Scientific Practice 'Bio-Techno-Logos'**
 Marta Bertolaso (ed.)

6 **Scientists' Expertise as Performance: Between State and Society, 1860–1960**
 Joris Vandendriessche, Evert Peeters and Kaat Wils (eds)

7 **Standardization in Measurement: Philosophical, Historical and Sociological Issues**
 Oliver Schlaudt and Lara Huber (eds)

8 **The Mysterious Science of the Sea, 1775–1943**
 Natascha Adamowsky

9 **Reasoning in Measurement**
 Nicola Mößner and Alfred Nordmann (eds)

10 **Research Objects in their Technological Setting**
 Bernadette Bensaude Vincent, Sacha Loeve, Alfred Nordmann and Astrid Schwarz (eds)

11 **Environments of Intelligence: From Natural Information to Artificial Interaction**
 Hajo Greif

Reasoning in Measurement

Edited by Nicola Mößner
and Alfred Nordmann

Routledge
Taylor & Francis Group

LONDON AND NEW YORK

First published 2017
by Routledge
2 Park Square, Milton Park, Abingdon, Oxon OX14 4RN

and by Routledge
711 Third Avenue, New York, NY 10017

First issued in paperback 2018

*Routledge is an imprint of the Taylor and Francis Group, an informa
business*

British Library Cataloguing in Publication Data
A catalogue record for this book is available from the British Library

Library of Congress Cataloging in Publication Data
A catalog record for this book has been requested

ISBN 13: 978-1-138-33194-5 (pbk)
ISBN 13: 978-1-8489-3602-7 (hbk)

Typeset in Times New Roman by
Servis Filmsetting Ltd, Stockport, Manchester

Contents

List of Figures   vii
Notes on Contributors   viii

1 Epistemological Dimensions of Measurement   1
   NICOLA MÖSSNER AND ALFRED NORDMANN

PART I
Founding Figures   **9**

2 Of Compass, Chain, and Sounding Line: Taking Thoreau's
   Measure   11
   LAURA DASSOW WALLS

3 Operationalism: Old Lessons and New Challenges   25
   HASOK CHANG

PART II
Images as Measurements   **39**

4 Photo Mensura   41
   PATRICK MAYNARD

5 The Media Aesthetics of Brain Imaging in Popular Science   57
   LIV HAUSKEN

6 Compressed Sensing—A New Mode of Measurement   73
   THOMAS VOGT

7 The Altered Image: Composite Figures and Evidential
   Reasoning with Mechanically Produced Images   87
   LAURA PERINI

8   Visual Data—Reasons to be Relied on?   99
    NICOLA MÖSSNER

9   Pictorial Evidence: On the Rightness of Pictures   111
    TOBIAS SCHÖTTLER

PART III
Measuring the Immeasurable   **131**

10 Measurement in Medicine and Beyond: Quality of Life,
    Blood Pressure, and Time   133
    LEAH MCCLIMANS

11 Measuring Intelligence Effectively: Psychometrics from a
    Philosophy of Technology Perspective   147
    ANDREAS KAMINSKI

12 The Klein Sexual Orientation Grid and the Measurement
    of Human Sexuality   157
    DONNA J. DRUCKER

13 The Desert and the Dendrograph: Place, Community,
    and Ecological Instrumentation   170
    EMILY K. BROCK

PART IV
Calibrating Mind and World   **187**

14 Scientific Measurement as Cognitive Integration: The Role
    of Cognitive Integration in the Growth of Scientific Knowledge   189
    GODFREY GUILLAUMIN

15 Measurements in the Engineering Sciences: An
    Epistemology of Producing Knowledge of Physical Phenomena   203
    MIEKE BOON

16 Uncertainty and Modeling in Seismology   220
    TERU MIYAKE

17 A Model-Based Epistemology of Measurement   233
    ERAN TAL

    Index   254

List of Figures

2.1 Thoreau's Survey of Walden Pond 12

4.1 Étienne-Jules Marey and his measuring devices 43

4.2 'Hurter & Driffield's Actinograph' slide-rule (photo Dick Lyon) 45

4.3 AP Photo/USA Track & Field photo finish 51

5.1 MRI scan: The view of the brain from above (transaxial), the profile (sagittal view) and the frontal (coronal view, with strongly marked neck bone) 63

6.1 Operation of a single pixel camera 78

6.2 Stokes's Quadrants 84

7.1 Composite MPIs (Mechanically Produced Images) paired with graphs 94

12.1 The Klein Sexual Orientation Grid (KSOG) by Fritz Klein 158

Notes on Contributors

Mieke Boon is a full professor of Philosophy of Science in Practice at the University of Twente, The Netherlands. She received her PhD (with honor) in Chemical Engineering from the Technical University of Delft, and studied Philosophy at the University of Leiden. Between 2003 and 2008, she has been working with a VIDI research grant from the Dutch National Science Foundation (NWO) on developing a Philosophy of Science for the Engineering Sciences. Currently, she continues this line of research on an NWO Aspasia grant. She published *Philosophy of Looking* (2009 in Dutch: *Filosofie van het Kijken: Kunst in Ander Perspectief*). She is a co-founder of the Society for Philosophy of Science in Practice (SPSP). Her publications can be found at the UT Repository http://www.utwente.nl/gw/wijsb/organization/boon/publication%20list. html.

Emily K. Brock is a research scholar at the Max Planck Institute for the History of Science in Berlin, Germany. She has a MS in Biology from the University of Oregon and a PhD in History of Science from Princeton University. She was a 2013 Fulbright U.S. Senior Research Scholar in the Philippines and a 2014 Fellow of the Rachel Carson Center for Environment and Society in Munich. Her book on the history of American forest management, *Money Trees: The Douglas Fir and American Forestry, 1900–1944*, is published in 2015 by Oregon State University Press. She is currently writing a book on the globalization of the Asian tropical hardwood industry.

Hasok Chang is the Hans Rausing Professor of History and Philosophy of Science at the University of Cambridge. He received his degrees from Caltech and Stanford, and has taught at University College London. He is the author of *Is Water H$_2$O? Evidence, Realism and Pluralism* (2012), and *Inventing Temperature: Measurement and Scientific Progress* (2004). He is a co-founder of the Society for Philosophy of Science in Practice (SPSP), and the Committee for Integrated History and Philosophy of Science.

Donna J. Drucker is instructor of English for special purposes at Technische Universität Darmstadt, Germany. She received a master of Library Science, master of arts in History, and a PhD in History from Indiana University, Bloomington. Her book *The Classification of Sex: Alfred Kinsey and the Organization of Knowledge*, was published in 2014 by the University of Pittsburgh Press, and it won the 2015 Bonnie and Vern L. Bullough Award from the Foundation for the Scientific Study of Sexuality. Her book *The Machines of Sex Research: Gender and the Politics of Identity, 1945–1985*, was published by Springer in 2014. She is now working on a history of non-hormonal contraceptive research in the 20th-century United States.

Godfrey Guillaumin is a full time professor for the Department of Philosophy at the Autonomous Metropolitan University (UAM), Mexico City, Mexico. He has a PhD in History and Philosophy of Science from the National Autonomous University of Mexico (UNAM). He was awarded a Fulbright-García Robles award for a postdoctoral research at the Center for Philosophy of Science at the University of Pittsburgh. His main areas of interest are the history of epistemology; history of scientific notions as evidence, measurement, and methodological rules; the problem of under-determination; the cognitive scientific progress, and the theory change. Both his current research projects focus on how epistemic ideas are gener-ated from development of measurement scientific practices, particularly in the history of ancient and modern astronomy, and how historical sciences develop their epistemic notions, mainly in evolutionary biology and palae-ontology. His books are *El surgimiento de la noción de evidencia* (UNAM, 2005), *Raíces metodológicas de la teoría de la evolución de Charles Darwin* (Anthropos, 2009), *Historia y estructura de La estructura* (UAM, 2012), and *Génesis de la medición celeste* (UAM/Tirant, 2016).

Liv Hausken, Dr. art., is Professor at the Department of Media and Communication at the University of Oslo and the head of the department's research area Media Aesthetics. Relevant publications in English include: 'The archival promise of the biometric passport', in I. Blom et al. (eds.), Memory in Motion: Archives, Technology and the Social (Amsterdam University Press 2016); 'The Visual Culture of Popular Brain Imaging', Leonardo: Journal of the International Society for the Arts, Sciences and Technology 48:1 (MIT Press 2015); 'The Normalization of Surveillance in Forensic Fiction', Nordicom Review 35:1 (2014): 3–16; *Thinking Media Aesthetics. Media Studies, Film Studies and the Arts* (ed., Peter Lang 2013); 'The Temporalities of the Narrative Slide Motion Film', in E. Røssaak (ed.), Between Stillness and Motion: Film, Photography, Algorithms (Amsterdam University Press 2011); 'The Aesthetics of X-ray Imaging', in A. Melberg (ed.), Aesthetics at Work (UniPub 2007); 'Textual Theory and Blind Spots in Media Studies', in M.-L. Ryan (ed.), Narrative across Media. The Languages of Storytelling (University of Nebraska, 2004).

Andreas Kaminski is head of the Department for Philosophy of Science and Technology at HLRS (University of Stuttgart). He studied Philosophy, and postdoc at the Technische Universität Darmstadt in Germany. He studied Philosophy, German Language and Literature, Sociology in Darmstadt and Berlin and received his PhD in Philosophy at the TU Darmstadt. His doctoral dissertation, titled *Technik als Erwartung. Grundzüge einer allgemeinen Technikphilosophie* (Bielefeld: Transcript 2010), conceptualizes technology as (different types of) expectations. Currently, he is working on epistemic opacity related to computer simulations and machine learning algorithms, furthermore on a dialectic theory of trust and testimony. He is one of the editors of the *Jahrbuch Technikphilosophie* and speaker of the DFG Scientific Network *Geschichte der Prüfungstechniken 1900–2000*.

Patrick Maynard is a graduate of the University of Chicago and Cornell University, emeritus professor of Philosophy, Western University, Canada, and has also taught philosophical classics and philosophy of art at the universities of Michigan-Ann Arbor and California-Berkeley, as well as Simon Fraser University. He is author of *The Engine of Visualization: Thinking Through Photography* (Cornell University Press, 1997), and *Drawing Distinctions: The Varieties of Graphic Expression* (Cornell University Press, 2005), co-editor of the Oxford *Aesthetics* Reader, author of articles in many journals (mostly cited in *PhilPapers*), and has lectured widely. A native of Trinidad, citizen of the USA and Canada, he lives in England, practicing philosophical research and visual art.

Leah McClimans is an associate professor of Philosophy at the University of South Carolina. She completed her PhD in 2007 at the London School of Economics and did a Postdoc in Medical Ethics at the University of Toronto in 2006–2007. Her work then and now looks at the ethics and epistemology of measurement focusing on patient-reported outcome measures and the evaluation of healthcare more generally. Her articles have appeared in *Theoretical Medicine and Bioethics, Journal of Medicine and Philosophy, Bioethics* and *Quality of Life Research*.

Teru Miyake is an assistant professor of Philosophy at Nanyang Technological University in Singapore. He has a BS in Applied Physics from Caltech and a PhD in Philosophy from Stanford University. His area of specialty is the philosophy of science, and much of his work has been about a special kind of scientific problem—that of trying to acquire knowledge about complicated systems to which we have limited access. He has done work on planetary astronomy, particularly on Kepler and Newton, and he has recently been working on 20th-century geophysics. He has published papers on these topics in *Philosophy of Science* and *Studies in History and Philosophy of Science*.

Nicola Mößner is a Junior Fellow at Alfried Krupp Wissenschaftskolleg Greifswald in Germany. She received her MA in German Literature and

Linguistics at the University of Hamburg and her PhD in Philosophy at the University of Münster. Her thesis is about the epistemology of testimony and the special case of media reports, published as *Wissen aus dem Zeugnis anderer – der Sonderfall medialer Berichterstattung* (Paderborn: mentis 2010). She edited (together with Dimitri Liebsch) *Visualisierung und Erkenntnis – Bildverstehen und Bildverwenden in Natur- und Geisteswissenschaften* (Cologne: Herbert von Halem, 2012). Currently she is writing a book about conceptual and epistemological questions concerning visual representations in science. Her main research interests comprise philosophy of science and social epistemology.

Alfred Nordmann is professor of Philosophy at the Technische Universität Darmstadt and adjunct professor at the University of South Carolina, Columbia. His interests in the philosophy of science concern the formation and contestation of fields of inquiry such as chemistry and theories of electricity in the 18th century, mechanics, evolutionary biology, and sociology in the 19th century. In particular, he sought to articulate implicit concepts of science and objectivity. In 2000, he embarked on a similar endeavor in regard to nanoscience and converging technologies which has led him to promote and develop a comprehensive philosophy of technoscience. Since the technosciences require new answers to the familiar questions of knowledge and objectivity, theory and evidence, explanation and validation, representation and experimentation, he is seeking to address these and related questions in his current work.

Laura Perini is an associate professor of Philosophy; Coordinator of Science, Technology and Society (STS), and Chair of Philosophy at the Philosophy Department at Pomona College. She received her MA in Biology from the University of California, Los Angeles, and her PhD in Philosophy from the University of California, San Diego. Her research focuses on the use of images in science, particularly in the life sciences. She has published a number of articles analyzing the relationships between the semiotics of scientific images and their epistemic role, including papers in *Philosophy of Science, International Studies in the Philosophy of Science, Biology and Philosophy*, and *The Journal of Aesthetics and Art Criticism*.

Tobias Schöttler works as a project manager and concepter at a creative agency. He received his MA in Philosophy, German Literature and Media Studies at the Ruhr-University Bochum and his PhD in Philosophy at the same university. His thesis deals with theories of representations, published as *Von der Darstellungsmetaphysik zur Darstellungspragmatik: Eine historisch-systematische Untersuchung von Platon bis Davidson* (Münster: mentis 2012). His main research interests comprise semiotics (especially picture theory and philosophy of language), epistemology, history, and philosophy of science. Currently, he is working on a research project concerning visual arguments in the sciences.

Eran Tal is an assistant professor at McGill University. He received an MA in History and Philosophy of Science from Tel Aviv University in 2006 and a PhD in Philosophy from the University of Toronto in 2012. His work develops a model-based epistemology of measurement that highlights the roles of idealization, abstraction, and prediction in measurement, and analyzes the inferential structure of measurement and calibration procedures performed by contemporary standardization bureaus. His articles have appeared in *The British Journal for the Philosophy of Science, Philosophy of Science, Synthese and Philosophy Compass*. He is the author of the entry "Measurement in Science" in the Stanford Encyclopedia of Philosophy.

Thomas Vogt is the Educational Foundation Distinguished Professor of Chemistry and Biochemistry, Adjunct Professor in the Department of Philosophy, and Director of the NanoCenter at the University of South Carolina. His research focuses on the structural characterization of new materials using x-rays, neutrons, and electron scattering and direct imaging using electron microscopy. His philosophical interests are in the history and philosophy of science, epistemology of measurements, and philosophy of chemistry. He is a Fellow of the Institute of Advanced Study at Durham University, the American Physical Society and the American Association for the Advancement of Science.

Laura Dassow Walls is the William P. and Hazel B. White Professor of English and a faculty affiliate in History and Philosophy of Science at the University of Notre Dame. Her books include *Seeing New Worlds: Henry David Thoreau and Nineteenth-Century Natural Science*; *Emerson's Life in Science: The Culture of Truth* and, most recently, *Seeing New Worlds: Alexander von Humboldt and the Shaping of America*, awarded the Merle Curti Prize in Intellectual History by the Organization of American Historians and the James Lowell Prize by the Modern Language Association. Currently, she is writing a biography of Thoreau (*Henry David Thoreau: A Life*, to be published by the University of Chicago Press in July 2017).

1 Epistemological Dimensions of Measurement

Nicola Mößner and Alfred Nordmann

This collection of chapters is dedicated to the epistemology of measurement. As such, it marks a clear distinction from older debates in measurement theory. By the same token it advances a renewed interest in measurement as a key element of research practice. To the extent that it has been interested in epistemology at all, *measurement theory* has generally been concerned with questions of knowability. Norman R. Campbell (1920) famously asked what things have to be like in order to be measurable on a certain scale, and inversely, what we know about the world when we can measure some but not all things. Equally famously, Stanley S. Stevens (1946) started from the practice of assigning numbers to objects or events and left questions of metaphysics and epistemology to be answered implicitly through the advance of quantification. This is the point where measurement theory, properly speaking, entered in to determine the axiomatic conditions that need to be satisfied so that domains of objects become amenable to measurement (see Suppes et al. 2007).

The epistemology of measurement, as we understand it, asks how measurement is accomplished. When do physical manipulations and associated modeling practices become sufficiently robust to afford something like an instrument or a measuring device, and how are measurements obtained from questionnaires, recording devices or instrument readings? By focusing on measurements as the hard-won results of conceptual as well as technical operations we no longer presuppose that measurement is always and exclusively a means of gaining knowledge of some target object or entity. Measurement may also provide knowledge about the degree to which things have been standardized or harmonized—it is an indicator of how closely human practices are attuned to each other and the world.

If this sounds somewhat abstract, the opening chapter by a literary scholar helps concretize it. *Laura Dassow Walls* contrasts two ways of thinking about measurement that are embodied in two philosophical writers often "lumped together" as American Transcendentalists. Ralph Waldo Emerson adopts a rather classical contemplative view whereby measurement is required to bridge the abyss between the world of the mind and the world of natural facts—how can the mind come to know the world if not by way of measurement? Henry David Thoreau seeks truth by engaging in the work of commensuration, that is, by contriving protocols for seeking out a common metre. This he achieves by adapting the sensory apparatus to the phenomena and by preparing devices and phenomena for the measurement process.

With Thoreau and Emerson in mind and their different ways of seeking out truth, the concept of representation cannot be presupposed as the unquestioned point of departure for reflections on measurement. Instead, one has to ask how representations are achieved—and when asking that question, one discovers that it is not always representation that is brought about when measurements are taken and when agreements are forged on measured values.

In order not to get ahead of ourselves and miss an important step in the analysis by assuming that measurements represent properties of phenomena that exist in the world, we need to find some other conceptual starting point.

In his chapter on Percy Bridgman's operationalism, *Hasok Chang* suggests a semantic conception according to which the meaningfulness of measurements derives from the coordination of different operations for measuring the same thing. It is through this multiplicity of meanings that measurements can be connected to the world and enable scientists to learn about reality.

There may well be other founding figures for an epistemology of measurement. As it stands, however, concepts such as *commensuration*, *operationalism*, and *active realism* open the door on a series of chapters that explore in terms that are not exclusively conceptual how measurements are achieved. Indeed, the chapter containing an analysis of photography as a technology of measurement is a prime example of this fruitful shift in perspective. *Patrick Maynard* has pointed out that theories of photography are ill-served if they begin with the photographic image, with depiction or representation, that is, with the pictures that we put in our photo albums or see in the newspapers. By beginning with these we posit a kind of original relation or paradigm, and view everything else in relation to it as variations on the original theme. When we start from the familiar photographic image, we treat photography as a kind of seeing, asking how photographic images are similar to or differ from retinal images, straining notions of receptive or passive seeing on the one hand and of interpretive or constructive seeing on the other. We are better served by viewing photography as a "family of technologies" that serve, for instance, the purpose of measuring and recording the intensity of light (see Maynard 2000: 3).

From this perspective, the application of this technology in the range of visible light is a special case, and within that range the creation of images that appear realistic to the sensory apparatus of humans is an even more special case—not unlike the technical decision to add visualization software to numerical simulations in order to render an output that resolves data in a visually familiar way. And in the course of making these technical decisions we can talk about causality, about rules for translation or transformation, and about robustness, such that we finally arrive at questions of veracity and can investigate notions of representational accuracy.

Only at first glance, then, does it appear puzzling that the topic of *images* is given a prominent place in a book about the epistemology of measurement. Especially in the context of today's big data sciences, data visualizations are more often than not the direct output of measurement processes. Providing cognitive access to huge amounts of data, it is the image that, in large parts of modern science, most clearly represents the epistemic virtues of comparability and objectivity commonly associated with the practice of measurement. Strategies for the management and visualization of data sets are discussed by the authors whose chapters appear in the second section of this book.

Liv Hausken challenges the implicit assumption of immediate visual access to the content of visualizations. She points out that even in the highly esoteric context of neuroscience, media experiences, and cultural background shape the perception of fMR images. To speak simply of *seeing* the information embedded in a brain scan would fail to describe correctly what exactly our epistemic practice of understanding scientific visualizations is about in this context.

A completely different strategy is presented by *Thomas Vogt*. Instead of compressing data after they have been collected, as in the case of the composite images discussed by Hausken, one might surrender the idea of acquiring every last bit (or byte) of data available and instead compress the data at the point of sensing, recording or collecting them. At first sight, this may raise concerns about a loss of information in the corresponding visualization. Contrary to this initial worry, Vogt highlights the fact that compressed sensing, as applied for example in the JPEG format of digital images, not only allows for complete reconstruction from a sparse data set but is often also the only available method in data-intensive research. Moreover, additional considerations—such as health risks due to long scanning procedures in X-ray diagnoses—speak in favour of compressed sensing.

There are other ways apart from data visualization, however, in which images are relevant for an epistemology of measurement. In quite a few instances, the technology of measurement itself involves *imaging technologies*. Here, instead of visualizing data after measurements have been performed, the initial measurement itself takes the form of a visualization. Patrick Maynard, for example, analyzes photography as a technology of detection; and Laura Perini discusses autoradiography as an analytic tool in biology. Here, images as measurements serve as surrogates for objects of research, allowing further manipulations and investigations.

A third dimension of images in the epistemology of measurement is their role in contexts of justification and the transmission of knowledge.[1] Measurement data derive their epistemic significance from serving as *evidence* in these contexts. In particular, the causal connection between detecting device and observed specimen is invoked to guarantee the reliability of measurement data and, thus, their ability to confirm or refute a theory or hypothesis. The same consideration applies to images if they are to serve as evidence. *Laura Perini*, however, questions this line of reasoning by pointing to the fact that images are often altered in publications to accommodate the assumed background knowledge of the audience. In the course of this, the causal connection between source and data is de-emphasized, if not sacrificed, for the sake of images that are better comprehensible in the communication of scientifically relevant information. This difficulty of balancing the epistemic value of presenting genuine data against the virtue of transforming them into measurements meaningful to human observers shows up not only in communication processes but in research as well. Thus, mechanically produced images are deeply enmeshed in problems of interpretation.

In view of these difficulties, *Nicola Mößner* scrutinizes the epistemic rationale for relying on visual data as measurement results. This rationale differs with respect to the aforementioned contexts, namely that of research and that of communicating results. By keeping the different requirements for each of these activities in mind, it is possible to avoid apparent difficulties associated with the epistemic role of images in science. The possibility of altering images in publications, for example, poses no great threat to their general evidential role, as it is the communicative act *as a whole* that has to be taken into account. Here, it is not the isolated interpretation of a particular image that accounts for the epistemic virtue of information transmission, but the author's credibility—her reputation as a scientist.

Finally, *Tobias Schöttler* discusses the rightness of images that are to serve as evidence in the context of measurement. There are two possible strategies for countering outright skepticism vis-à-vis our ability to assess this rightness. Schöttler rejects as unsatisfactory both the naturalist strategy of referring to causal processes and the conventionalist strategy of determining the referent of an image relative to a given system of symbols. Instead, he proposes a pragmatist approach according to which one begins by assuming the rightness of pictures and then probes that assumption by subjecting it to a local skepticism. The measurements themselves thus guide the process of fine-tuning the initial assumption and identifying precisely what it is that is being measured.

The pragmatist strategy becomes especially salient when we seek to measure things which cannot be accessed in straightforward ways independently of the measuring process itself. Intelligence, sexual orientation, and quality of life, for example, are subject to attempts to *measure the immeasurable—*objects that appear ephemeral and elusive and that become well defined or real only to the extent and in the course of being measured.

Such attempts to measure the seemingly immeasurable are discussed by the authors who have contributed to the third section of this book. *Leah McClimans* begins by considering concerns about the measurability of quality of life. Comparing the difficulties of measuring quality of life to the relative ease of measuring blood pressure, she points out that the key difference between these cases is not the lack of a criterion or gold standard against which to assess the obtained measures. Rather, it is the lack of a theory that would allow quality-of-life researchers to know what they are measuring and whether their measurement procedures are adequate.

Andreas Kaminski addresses questions around whether *intelligence* is measurable and whether psychometrics yields informative results. Even in the absence of a decisive answer to these questions, however, psychometrics can be said to be effective in that the measurement of intelligence influences its object, namely, the individuals who are being assessed and their behaviors. Having one's intelligence tested and adopting such tests as scales for personal improvement reveals the subject's interest in learning more about themselves and taking charge of their own identity.

Such an interest in the self is also highlighted by *Donna J. Drucker* who looks at the domain of sex research. She considers the history of attempts to determine objective criteria for measuring *sexual orientation* by way of questionnaires. Having started out in the context of academic research, a proliferation of such questionnaires now circulates on the Internet. Here, the measurement tool serves to stimulate contemplation of one's sexual lifestyle or self. While this would appear to be a good thing (regardless of the veracity of the findings), participants in most instances are left to themselves and remain at a loss to interpret the results.

Finally in this section, *Emily K. Brock* reminds us that the emergence of new domains of application for measuring practices is often driven by the *technicians* responsible for inventing or developing measuring devices. She describes the invention of the "dendrograph", an ecological measuring tool, by an instrument-maker at the Carnegie Desert Lab. This device affords the measurement of a tree's minute physiological reactions to environmental changes. The availability of this new instrument was intended to advance *ecology* as a genuinely scientific discipline. This relatively new disciplinary field did indeed come to enjoy an enhanced scientific reputation once the collection of data in the field had become as precise as laboratory measurements thanks to the new instrument.

The last part of this book is dedicated to practices of reasoning in measurement. Understanding measurement correctly in an epistemological sense implies consideration not only of its representational and technological dimensions, as discussed so far, but also of the requisite *calibrations of mind and world*. In other words, ways of thinking about nature and its phenomena are continually being attuned to practices of gaining knowledge about them.

In this context, *Godfrey Guillaumin* discusses how measurement is implicated in the advancement of science. One difference between Ptolemy and Copernicus lies in the way each of them explains the phenomena. But the transformation of our understanding of nature effected by the so-called Copernican Revolution comes by way of Kepler. This is due to the cognitive integration performed in the very act of measurement which, after Kepler, affords physical meaning in novel ways.

The dependence of measurement upon the overall system of concepts and practices in which it is embedded is also emphasized by *Mieke Boon*. From the perspective of engineering science, measurement appears as an integral component of design practices. Engineers create new technologies and therefore require knowledge of the physical-technological environment responsible for producing the relevant effects. In this sense, Boon's approach highlights the constructive, if not creative, aspect of measurement.

Another way of calibrating mind and world is presented by the last two authors in this section, both of whom engage issues of modeling. *Teru Miyake* discusses the seismological practice of using earth models to infer properties of the interior of our planet. Such models, he argues, are best understood as instruments of measurement. This approach leads toward an understanding of the epistemology of measurement in which the creation and maintenance of such models becomes a constructive endeavor—one which transforms a highly inferential representation into an instrument or tool for exploration.

Eran Tal's analysis puts the final touch to the volume by explicitly articulating the contours of an epistemology of measurement. His chapter thus simultaneously provides a summary of many points from the preceding chapters and an outlook towards future work. In particular, he puts forward his view of calibration as consisting of comparisons between models of measurement processes. Here again, we are offered the productive insight that measurements sometimes play a representational role but always involve calibrations of mind and world.

Acknowledgments

This book grew out of the conference *Dimensions of Measurement* at the Center for Interdisciplinary Research (ZIF) in Bielefeld (Germany), in March 2013—as did its complementary volume *Standardization in Measurement* (edited by Oliver Schlaudt and Lara Huber, Routledge 2015). We would like to thank Kathleen Cross for editorial support. Nicola Mößner's work on this volume was supported by the German Research Foundation (DFG) through its funding of the project *Visualisierungen in den Wissenschaften—eine wissenschaftstheoretische Untersuchung* (MO 2343/1–1).

Note

1 Klaus Hentschel (2014) states that: "[t]oday about one-third of the available space on the pages of leading scientific journals is devoted to nontextual material, be it glossy photographs, plots of functional dependencies of variables or schematic diagrams" (ibid.: 43).

References

Campbell, N. R. 1920. *Physics: The Elements*. Cambridge: Cambridge University Press.

Hentschel, K. 2014. *Visual Cultures in Science and Technology. A Comparative History*. Oxford: Oxford University Press.

Maynard, P. 2000. *The Engine of Visualization—Thinking Through Photography*. Ithaca, NY: Cornell University Press.

Stevens, S. S. 1946. On the Theory of Scales of Measurement. *Science* 103:*2684*: 677–80.

Suppes, P., Luce, R. D., Krantz, D., and Tversky, A. 2007. *Foundations of Measurement*. Vols. 1–3. Mineola, NY: Dover Publications.

Part I

Founding Figures

2 Of Compass, Chain, and Sounding Line: Taking Thoreau's Measure

Laura Dassow Walls

It still seems a truth universally admitted that a romantic poet cannot carry a tape-measure in his pocket. In Ralph Waldo Emerson's words, the "half-sight of science" fails to recognize that we do not learn "by any addition or subtraction or other comparison of known quantities" but only by "untaught sallies of the spirit" (Emerson 1983: 45, 43). What to make, then, of his friend Thoreau—whose most "romantic" book, *A Week on the Concord and Merrimack Rivers*, contains a hymn to comparisons of known quantities? "How many new relations a foot-rule alone will reveal … What wonderful discoveries have been, and may still be, made, with a plumb-line, a level, a surveyor's compass, a thermometer, or a barometer!" (1980: 363). Who was distinguished, as Emerson remarked, for his "natural skill for mensuration," who was constantly ascertaining the height of trees and mountains, "the depth and extent of ponds and rivers," and who "could pace sixteen rods more accurately than another man could measure them with rod and chain" (Emerson 2008: 395, 399)? Who notched his walking stick with inch-marks and recommended that on every excursion one should carry a "compass; plant-book and red blotting-paper … small pocket spyglass for birds, pocket microscope, *tape-measure*, [and] insect-boxes?" (Thoreau 1972: 319). There was, evidently, at least *one* romantic poet with a tape-measure in his pocket.

Emerson further concluded that Thoreau's peculiar gift for mathematics plus his intimate knowledge of the Concord landscape "made him drift into the profession of land-surveyor" (2008: 395). Here, too, he misses the point: Thoreau was hardly a poet who "drifted" into a boring day-job to pay the bills. The very inception of *Walden*, his greatest work of art, was a survey map so meticulous its accuracy has been verified by modern instruments (Figure 2.1).

This is a highly professional survey of its eponymous subject, based on a physically arduous set of measurements Thoreau took during his first winter in residence at the Pond. *Walden* was published in 1854, the height of America's infatuation with that floating temple of mensuration, the scientific exploring expedition. Some readers thought this, its sole illustration, was a joke, the punchline of Thoreau's "caricature of the Coast Surveys" (Thoreau

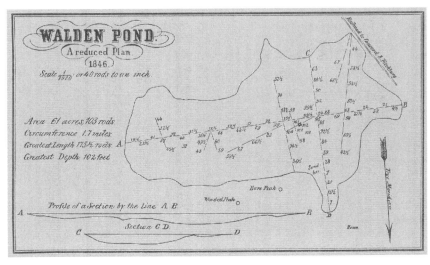

Figure 2.1 Thoreau's Survey of Walden Pond.
Source: H. D. Thoreau, Walden (1854).

1906: 102–3). Wrong again—although at least such a reading acknowledges Thoreau's map as part of the text; many modern editions, shamefully, leave it out altogether, assuming that Thoreau's iconographic act of mensuration can have no place in literary art.

On the contrary: mensuration is the heart of Thoreau's literary art; it is the soul of his poetics; it is the hand by which he grasps the world. Critics who marginalize his Walden survey fall in line with the lamentable Romantic tradition holding that poetry is blinded by fact; that the more closely Thoreau retraced his steps around Walden—measuring stream depths, counting tree rings, recording the leafing and blooming and leaf-fall of every plant, the arrival date of every bird, the days when he donned his gloves, or his neighbours doffed their coats—the farther he fell from grace. Then how can we explain that the more closely Thoreau followed his genius for mensuration, the more original he became? As Stanley Cavell asserts, "The human imagination is released by fact" (1992: 75). The resources of measurement allowed Thoreau to resist Emerson's version of "Nature" and work out, instead, a new poetics of science that would give, not "half-sight," but *full* sight.

The differences between Emerson and Thoreau are diagnostic. Each of them starts with the same problem: human alienation from nature, or the fall from faith to skepticism. As Emerson wrote in *Nature*, the touchstone of American Romanticism: "We are as much strangers in nature, as we are aliens from God. We do not understand the notes of birds. The fox and the deer run away from us; the bear and tiger rend us." In this, our fallen state, nature becomes "a fixed point whereby we may *measure* our departure" (1983: 42). Emerson's solution is to construct a system of analogical relationships between man and nature: "the whole of nature is a metaphor of the human mind. The laws of moral nature answer to those of matter as face to face in a glass ... The axioms of physics translate the laws of ethics" (ibid.: 24).

By thus delimiting nature and humanity as two great, mirror-imaging oppositions he sets in motion a metaphorical dynamic that never ceases to surprise and energize his poetics. In fact, Emerson's romance with duality is so persuasive that he became one of the great architects of American modernism: transcendent, inviolable Nature becomes the infinite fountainhead powering human creation of a second nature that draws on, and will ultimately eclipse altogether, the merely empirical phenomena of the material world, the disposable *body* of nature, which like the body of man is readily subdued and remade by human will. All this may be done without violating what really matters, the stainless ideal of natural law. Everywhere in Emerson this axiom of self-similarity reigns unchallenged; in his universe once one needs no protocols of translation, for when the world turns to glass and nature's laws shine through, one needs no mediation—one has become no less than God in nature.

Thoreau began his creative life as Emerson's protégé, and it took well over a decade for him to break away from Emerson's poetics and find his own, original, voice. He is popularly misunderstood to have been a misanthrope whose closeness to nature marked his alienation from humanity. In fact, he helped support a large household, and while he occasionally took his famous walks alone, he frequently walked with a companion and stopped to chat with the farmers working their fields. The moment of his most profound alienation occurred not at Walden Pond but near the top of Katahdin, Maine's highest and most rugged mountain. Having left his friends behind to strike for the summit, he found himself alone in a wilderness of rock and cloud, where for the first time he confronted "no man's garden, but the unhandselled globe ... Matter, vast, titanic." His profound sense estrangement from his own body— "I stand in awe of my body, this matter to which I am bound has become so strange to me ... What is this Titan that has possession of me?"—finally left him howling: "*Contact! Contact! Who* are we? *Where* are we?" (Thoreau 1972: 71).

His response to this crisis was to return to his house at Walden Pond on the outskirts of town, where he studied how to multiply means and technologies of "contact" with "Matter" itself, that incommensurable Titan to which he had discovered himself so inextricably bound. Over the next few years Thoreau became a skilled naturalist, a field collector for Louis Agassiz's natural history museum at Harvard, a serious and increasingly competent amateur botanist, and an ethnographic researcher. As respect for him grew, Boston's scientific elite appointed him a corresponding member of the Boston Society of Natural History and invited him to join the American Association for the Advancement of Science, all as he was nurturing and polishing a shelf of literary classics.

The metaphysical crisis Thoreau experienced on Katahdin was conditioned by the decline of religious faith (Thoreau was a religious skeptic) and the rise of a scientific world view that showed the earth to be unfathomably old, the universe to be infinitely large. One could say that commensuration was the primary problem of the age: as new measures showed both the unimaginable age of the rocks and the immense distance of the stars, the human mind was, abruptly, confronting an incommensurable universe that failed, as Charles Taylor has observed, "to touch bottom anywhere" (1997: 325). Where Emerson's solution was to construct a cosmos erected on mirror-image analogies, Thoreau's was to construct protocols of commensuration derived from the emerging conventions of global exploration science. He followed his new protocols out and back from Walden Pond, interlacing across the planet a vast network of relations: a multiple world, turbulent, contingent, bottomless, calibrated into meaning by way of pencils and paper, foot-rules and measuring tapes, compass, surveyor's chain, and sounding line.

The World in Three Dimensions, I: The Metrics of Poetry

Early in his thinking about science, Thoreau offered the insight that "Science applies a finite rule to the infinite—& is what you can weigh and measure and bring away" (1981: 44). Thus he acknowledged the institutional freight of science and its function as a means of circulation, anticipating Bruno Latour's discussion of "measure" as constructing "a commensurability that did not exist before their own calibration" (Latour 1993: 113). Thoreau's insight does not divorce science from poetics, but identifies the very ligature that connects them: the word "measure" is allied to the Greek word *mētis*, wisdom or skill, and to its sister word *metron*, which includes not only measure as length or rule but also proportion, as in "measured" language (which is calculated, careful, or restrained), and poetic *meter*, a measured or rhythmic arrangement of words.

Latour further remarks that "Worlds appear commensurable or incommensurable only to those who cling to measured measures. Yet all measures, in hard and soft science alike, are also measuring measures" (ibid.). This too points to the ligature connecting science and poetry, as when Emerson protests that true poets are not those who cling to "industry and skill in metre"— what one could call *measured* measures—but rather those who utter "the necessary and causal," who do not merely sing but are "children of music:" "For it is not metres, but a metre-making argument, that makes a poem,—a thought so passionate and alive, that ... it has an architecture of its own, and adorns nature with a new thing" (1983: 450).

Emerson's call for "a metre-making argument," a *measuring* measure organically related to the structure of the universe, famously inspired Walt Whitman to abandon "measured measures" and devise his own innovative metrics; it also moved a young Thoreau to articulate his own theory of measure: "To the sensitive soul, the universe has its own fixed measure, which is its measure also ... When the body marches to the measure of the soul, then is true courage and invincible strength" (Thoreau 1981: 96). There is a whiff of Emerson's dualism here, as the aspiring soul strives to reach the fixed measure of the universe, but more remarkable is the involvement of the body, as it "marches" to both the soul and to the universe, lining up all three—body, soul and universe—into cosmic harmony. A decade later, Thoreau expanded this journal passage into a meditation on health, defined as the state of being in which "the pulse of the hero beats in unison with the pulse of Nature, and he steps to the *measure* of the universe" (1980: 175). Translated into *Walden*, this becomes one of Thoreau's most famous lines: "If a man does not keep pace with his companions, perhaps it is because he hears a different drummer. Let him step to the music which he hears, however *measured* or far away" (1971: 360). Thoreau has by now fully switched from a "measured measure"—the drumbeat which commands our steps in universal unison—to a "*measuring* measure," the "music" which calls each of us to respond with our own "steps." The almost universal misquoting of this passage—"let him MARCH to the music"—forces Thoreau's language back to the military image which he had left behind, as if we still require the universal measure, the lockstep enforcer, the drumbeat of war. But Thoreau's mature language points toward peace, not war—toward the footfalls of a walker or the steps of a dancer.

Stepping to the measure of a different music evokes as well Thoreau's practice of walking not as exercise but as meditation, a bodily practice so central to his spiritual discipline that he spent a large part of virtually every day on foot, walking. The rhythm of his walking body generated the measure of his prose—quite literally; as he observed, the length of his daily walks measured the length of his daily journal. This acknowledges what every hiker knows, how the body thinks *through measure*, the step or stride forming the metrical unit that collocates body, soul and universe by sound and rhythm, commensuration not merely by "contact" but by physical immersion. Nature itself thinks through meter: Thoreau relishes the cycle of the seasons as a "rhyme" that "nature never tires of repeating" (1906: 168), and around this figure he built most of his later work. But, lacking metrical models that wouldn't be available until the twentieth century, Thoreau abandoned formal poetry to invent a "free verse" prose that allowed him to step to the measures he heard, however far away from literary form.

The World in Three Dimensions, II: The Metrics of Science

Given that walking and thinking should measure each other, the genre that attracted Thoreau was the "Excursion," the round-trip outing there and back again, cast in the form of an "essay," trial-run or attempt. Thoreau modeled his excursive writings on the exploration narratives of such scientists as Alexander von Humboldt and Charles Darwin. Mobility lay at the heart of his method, for mobility allowed him to structure his work along trajectories and transepts, creating a matrix for comparative methods and, importantly, for recursion: there and back again, and again, and again. Professional surveying was essential to this recursive process, for it enabled Thoreau to cross private property boundaries without challenge and without harm—taking, as he once joked with a lecture audience, "a surveyor's and a naturalist's liberty," according to which "I have been in the habit of going across your lots much oftener than is usual, as many of you, perhaps to your sorrow, are aware" (Thoreau 2007: 166).

Thus his method, practiced daily for over a decade, was to go over the same ground repeatedly, building up a vast data set of comparisons which allowed him to detect and measure differences. Similarly, his method as a writer was to lay down tracks for the reader to follow, traces that could be tracked there and back by anyone. Hence his writings are all topographic, centered on a *topos*, a topic, as place-centered; hence also, they are all cartographic, deeply invested in mapping, whether creating cartographic surfaces for imaginative play or tracing and invoking actual maps.

Thoreau's inspiration for this mode of writing on and with the land— "geo-graphy," earth-writing—was the age's great physical geographer Alexander von Humboldt, whose major writings became available to Thoreau late in the 1840s, exactly when he was seeking to escape Emerson's influence and strike out definitively on his own. Indeed, Thoreau opened his first "excursion," "A Walk to Wachusett," by identifying his felicitous pun with Humboldt: by means of the misty mountains on the New England horizon he could translate the world's great "poets and travelers," whether he sat with the classical Homer, roamed with Virgil, or "with Humboldt *measured* the more modern Andes and Teneriffe" (2007: 29). In his first book, Thoreau paid tribute to the protocols of measurement set by the "more modern" Humboldt:

> The process of discovery is very simple. An unwearied and systematic application of known laws to nature, causes the unknown to reveal themselves. Almost any mode of observation will be successful at last, for what is most wanted is method. Only let something be determined and fixed around which observation may rally. How many new relations a foot-rule alone will reveal, and to how many things still this has not been applied! What wonderful discoveries have been, and may still be, made, with a plumb-line, a level, a surveyor's compass, a thermometer, or a barometer! Where there is an observatory and a telescope, we expect that any eyes will see new worlds at once.
>
> (1980: 363)

Twenty years ago, this passage inspired the title of my book on Thoreau and science, *Seeing New Worlds*. There I proposed that Thoreau derived his new "method" directly from "Humboldtian science," borrowing the term from Susan Faye Cannon, the historian of science who defined it as "the accurate, measured study of widespread but interconnected real phenomena in order to find a definite law and a dynamical cause"—which was, she adds, for some decades "the major concern of professional science," its leading edge and avant-garde (1978: 76–7, 105). In my own, more literary explication, I suggested that Humboldtian science could be outlined in four imperatives: Explore, Collect, Measure, Connect. I further suggested that its goal was an *"empirical* holism" which, in contrast to the tradition of "rational holism" followed by Emerson, approaches the whole through study of the myriad interconnections of its constituent and individual parts (see Walls 1995: 134, 60–93; Walls 2009: 262–8).

This mnemonic remains useful, for it encapsulates the method of *field-work*: by exploring, the self mobilizes along transepts or passages, collecting objects and inscriptions and measurements along the way and bringing them all back to a central point to be compared and connected. In recent years, much new work has been done to extend our understanding of Thoreau's protocols of measurement. What we are discovering is an empirical ecopoet-ics that is deeper, stronger and stranger than hitherto suspected, pointing directly to a powerful connection between poetics and science.[1]

I would argue, then, that constructing our understanding of Thoreau in ignorance of this vast bank of information points not to Thoreau's "half-sight" but to our own. This ignorance is itself worth investigating, but instead I would like to turn to the related question of "nescience," or "non-knowledge" as itself a productive force in generating the wonder and curiosity that are preconditions for inquiry. Starting with Thoreau's amaze-ment that measure constructs relations that allow us to "see new worlds," I would suggest that the pivotal point which our ignorance has obscured is the nature of "measure" as the core of a distinct, and extremely fruitful, scientific-poetic method. Confronted by a "bottomless" world newly plunged into cosmic incommensurability, Thoreau sought to imagine humankind not as, in Emerson's words, "a stupendous antagonism, a dragging together of the poles of the Universe" (1983: 953), but rather as a walker across the face of the planet, "measuring" it with mind embodied, crossing by small steps the chasms which no longer yawned into the epistemological abyss of Being but resolved into a succession of foot-bridges, small links constructing an archipelago that would grow both Being and Knowledge from the center out, joining with other archipelagos to create a network of translational chains. As Latour would point out, this is not some Utopian fantasy but a pretty good description of where we actually are, living here in the Middle Kingdom between those great antagonistic Poles of the Universe, if only we were able to declare "peace" instead of war and pay closer attention to the things that constitute our world and all the forces and beings those things gather together (see Latour 1999: 69–73).

Thoreau's basic protocol was the notebook. Almost none of his penciled "field notes" have survived, for after expanding them into his journal volumes he tossed them into the fire. But he did leave behind dozens of notebooks, including his journal of over 2,000,000 words, out of which he crafted his published essays and books. At his death he also left behind scores of charts on which he had been collating thousands of seasonal observations, forming the working index for his last two books, *Dispersion of Seeds* and *Wild Fruits*, literary experiments that, while unfinished, nevertheless contain some of his finest writing.

Such "nonliterary" writings have been ignored by generations of literary critics attuned to polished, formal fictions. But today the question presses: How can we read such "unreadable" paratexts? Here science studies offer some help: one can read them, in Hans-Jörg Rheinberger's words, as "epistemic things," "graphematic traces" that are of the order, not of printed literature, but of "experimental engagement and entanglement." This means they belong to Thoreau's "knowledge regime," located on the interface between "the materialities of experimental systems"—Thoreau's daily practices of walking, collecting, measuring, testing—and the published "conceptual constructs" now sanctioned as great literature (Rheinberger 2010: 244–5).

As Rheinberger notes, the protocols of reducing lived experience to a textual surface facilitate ordering and rearranging data, including presenting sequential events in synchronic form (as in Thoreau's phenological charts), and presenting temporal relations as spatial (as in his Walden survey). What Rheinberger calls "condensation effects" become critically important for Thoreau: depending on how data are arranged and the extent to which they are compressed, new patterns can become visible, and transformations can be reversed to run in either direction. Hence a protocol can become "epistemically productive"—in effect, turning the laboratory (or the pond, forests and fields) "into a writing surface" which is not merely reductive, but generates "new resources and materials" that open whole new spaces, preventing the process of knowledge from shutting down prematurely (see ibid.: 245–6). This is the logic behind Thoreau's nescientic wonder that a mere foot-rule enables us to "see new worlds at once"—or to make mind, body, and universe, Emerson's vast and antagonistic incommensurables, commensurate.

The paradigmatic instance of this is Thoreau's Walden Pond survey (Figure 2.1). Late in his first book, Thoreau took his first public swing at Emerson's philosophy: "May we not *see* God? Are we to be put off and amused in this life, as it were with a mere allegory? Is not Nature, rightly read, that of which she is commonly take to be the symbol merely?" (Thoreau 1980: 382). The entirety of *Walden* develops this challenge, beginning with the original 1846 survey which did not simply measure and record the pond, but *generated it as a site of meaning*. Hence the finished 1854 figure must be read not as image but as event, one which produces the Pond as both a highly specific space, or *topos*, and a temporally-folded narrative. Here, on the flat plane of the page, Walden the pond becomes *Walden* the book, exactly here where the space-times of both intersect with those of the reader: here the railroad track, built just before Thoreau moved to the pond, cuts across the cove opposite the house he built, here, marked simply "House;" here the holes he cut in the ice to sink his sounding line; here "Bare Peak," its property owner-ship, like all property lines, erased; but the engraver's lines are present, here, in the tremor of the stylus as it plows the surface of the metal plate from which the map was printed.

It is important to resist the traditional notion that this figure "represents" the pond, for that cannot be the point of this oddly nonrepresentational drawing; if it were, why not simply provide a picture of the pond? Instead, Thoreau is attempting to "cast his argument in the form of a diagram," in a parallel to Darwin's famous diagram of the evolutionary Tree of Life, of which Gillian Beer has written: "[Darwin] did not simply adopt the image of a tree as a similitude or as a polemical counter to other organisations. He *came upon it* as he cast his argument in the form of diagram. This 'materialization' of the image is important in understanding its force for him. It was substan-tial, a condensation of real events, rather than a metaphor." Meaning, she says further of Darwin, does not exist separate from matter; it inheres "in activity and in relations" (1983: 38, 41).

How this might be so becomes even more clear as one looks at Thoreau's narrative of the "real events" that he condensed into the substance of his own diagram: "As I was desirous to recover the long lost bottom of the Walden Pond, I surveyed it carefully, before the ice broke up, early in '46, with com-pass and chain and sounding line." The text that follows interlaces details of his method with the meander of his thoughts, dwelling particularly on how he could fathom the pond's depth "easily with a cod-line and a stone weighing about a pound and a half, and could tell accurately when the stone left the bottom, by having to pull so much harder before the water got underneath to help me" (Thoreau 1971: 285, 287). Patrick Chura's recent explication of Thoreau's survey further details the complex materiality belied by this clean and apparently uncomplicated image.

Thoreau moved to the Pond in July 1845, but since soundings from a floating boat were not sufficiently accurate, he waited until the pond froze, in January 1846. Once the ice was strong enough to bear his weight, Thoreau spent at least a week out in the bitter cold, flagging some 20 or so sighting posts along the shoreline, then hauling heavy instruments—the compass on its tripod, the metal surveying chain, an axe, the cod-line and weight, the graduated staff (probably held by an assistant)—across the ice, on which he established a 925-foot baseline with two primary traverse stations, from which he conducted an angle intersection survey of the pond's perimeter, "measuring almost 2,900 feet in 66-foot increments." Then Thoreau cut, by hand, well over 100 individual holes through the ice, each time lowering the plumb-line into the water and hauling it up again. At each point he paused to note every bearing and measure with pencil on paper.

Finally, when Thoreau prepared the original 1846 survey for publication he simplified or "reduced" it by eliminating many of the soundings, reducing the scale and, interestingly, reorienting his compass and hence the angle of his baselines from magnetic north to his "True Meridian," a highly significant re-orientation to the North Pole which took him several days' painstaking work to ascertain (see Chura 2010: 30–44, 114–20). It was this "reduced plan" of 1854 that the engraver reproduced in professional style and formal lettering.

All of this is interesting enough, but Thoreau is just getting started: the "condensation" of physical data, important to both Rheinberger's and Beer's accounts, yields Thoreau the metaphysical pay-off by which his provisional fiction of measurement was stabilized into significance, into a *metaphor*:

> When I had mapped the pond by the scale of ten rods to an inch, and put down the soundings, more than a hundred in all, I observed this remarkable coincidence ... that the line of greatest length intersected the line of greatest breadth exactly at the point of greatest depth.
>
> (Thoreau 1971: 289)

Compression of data has revealed a new pattern, which Thoreau leaps to enlarge:

> What I have observed of the pond is no less true in ethics. It is the law of average ... draw lines through the length and breadth of the aggregate of a man's particular daily behaviors and waves of life into his coves and inlets, and where they intersect will be the height or depth of his character.
>
> (ibid.: 291)

This is, as Beer observes of analogy, shifty, revelatory, and precarious, aligned with magic; both authoritative and absurd, it enchants by its very fragility. Sure enough, Thoreau tried the magic again, measuring a second, nearby pond, and with evident satisfaction reports a similar result; and he invites us to try the magic ourselves by refusing to provide the confirming lines on his diagram, forcing the incredulous reader to perform the act of collation in their own copy of *Walden*. By all these actions the pond has been *produced*, not merely recorded, as both an intensely physical placial being and also as a moral agent, a substantive real presence with voice and agency. Walden Pond becomes an affectingly "deep" character who even revealed still conceals, whose headlands shelter hidden coves no matter where one stands.

Thoreau, by this sequence of measurement, substantiates his transformation of the pond from background to being, object to agent: "A lake is the landscape's most beautiful and expressive feature. It is earth's eye; looking into which the beholder *measures* the depth of his own nature" (ibid.: 186). The enchantment of the pond is not broken by measure but produced by it; by this measure the human is not fallen from nature but closer to it, even befriending, and befriended. The culmination of Thoreau's poetics is the moment when he can speak to the pond, as to a person, as a being he perceives without quite recognizing: "Walden, is it you?" (ibid.: 193).

The fiction that propels Thoreau into this inquiry is that Walden is said to be bottomless. This is the dispute he deliberately settles "with compass and chain and sounding line," for as he remarks, without sarcasm: "It is remarkable how long men will believe in the bottomlessness of a pond without taking the trouble to sound it" (ibid.: 285). This dispute drives not only his work with axe and cod-line, but also the book's overall quest to "work and wedge our feet downward" through the "mud and slush of opinion" to find "a hard bottom and rocks in place" on which one can set "a gauge ... a Realometer" to measure for future ages the depth of the day's "shams and appearances" (ibid.: 98). The downward pull of planetary gravity governs as well Thoreau's resolution "to live *deliberately*, to front only the essential facts of life" and publish "a true account of it in my next excursion" (ibid.: 90–1, emphasis added): here Thoreau pursues yet another trope of measure, *deliberare*, the weighing or balancing of one's life (rather than one's gold) in the scales of justice.

As for weighing one's gold, Thoreau deals with this in his opening chapter, "Economy," by borrowing the metrics of global trade to publish his own personal "account," or financial balance sheets, to the world—right down to the "$28.12½" he spent on materials to build his house (see ibid.: 49). The point of this relentless pursuit of the art of mensuration would thus seem to be, as Simon Schaffer has suggested, to settle disputes: this is why Thoreau takes out his tape-measure. But *does* his inquiry "settle?" Does he really find that "hard bottom and rocks in place" on which to mount that novel measuring instrument, a "Realometer?" Or, as he accumulates languages both ancient and modern, Western and Eastern, colonial and indigenous, human and nonhuman, poetic and scientific, does he let slip the boundaries that stabilize a unitary meaning at all?

Emerson staged a world that no longer anywhere touches bottom in order to counter with a dynamic polarity of immense antagonistic forces, and in doing so he became one of the premier architects of what we call the "modern," in Latour's sense of the term. In his long dispute with Emerson, Thoreau staged in reply not rock-hard ontological certainty but a newly indeterminate world, what Karen Barad has called a "relational ontology" (or what I called a "relational knowing"), which studies how specific practices, or protocols, embody specific material configurations and phenomena (see Barad 2003: 814; Walls 1995: 147). That is, Thoreau's exacting and material practice of measurement instructs him that no measure exists independently of the one who measures nor of the instrument of measure. Instead, measurement is an emergent property of the intra-action of all these components, all of which exert agency: the pond that collaborates by freezing; the cod-line that links, the stone that sinks, the mud that arrests the stone's fall and the water that conveys its rise; the human hand hauling up the sounding line with greater or lesser effort depending on whether the water helps, or doesn't. Only in the intra-action of their agencies does the pond emerge as something real and meaningful: Walden, is it you? Thus, Thoreau's actions enact not a Cartesian cut between subject and object, but a Baradian "agential cut" which both untangles Thoreau and pond and surveying equipment, such that everyone knows their distinct and separate role, while it re-tangles them into a phenomenon that "comes to matter," *twice over*: the pond, itself; and the book, itself: Two Waldens, each of them entirely real and solid, each of them stabilizing the opposite end of a long, but not infinite, chain of practices and agents.

The World in Three Dimensions, III: The Metrics of Nature

When Thoreau howls on the top of Katahdin for "*Contact! Contact!*," he feels the incommensurability of nature as a mortal threat: his sublime is not a pleasing frisson but sheer terror that, as we are bound and made by bodies, these very bodies will unbind and unmake us too. Boundaries in Thoreau always dance with limitation: against his hymn to the foot-rule he offers a hymn to "extravagance," life without bounds, the transgressing "wild"—a hymn inspired, shockingly, by the rotting corpse of a dead horse whose stench insists we account for the fatal challenge posed by living as a mortal body (see Thoreau 1971: 318). The performance of his acts of commensuration is made on these terms: in the end, our life is utterly incommensurate with our death. Yet we keep on living. This fact astonished Thoreau, who entered his vocation as a writer through the gateway of tragedy, and discovered during his long recovery from despair how measure could redeem the abyss of nihilism into a ground of meaning.

Thus he played constantly with the notion of a true, or natural, metrics: not just arbitrary foot-rules and plumb-lines but something further, a metrics beyond measure. Could there be true measures, as alive as the beat of a pulse? Surely yes: this conviction sent Thoreau to the sun as the true measure of the day, and he noted how every stick and tree inscribed a sun-dial like the noon-mark on the housewife's kitchen floor, still visible even as the mechanical clock chimed the morning hours, even as railroad time forced all the clocks of Concord to chime in synchrony not with the sun but with the new Standard Time, set for all the globe by the Greenwich Meridian. The Roman calendar may have organized his journal, but the seasonal cycles organized his writings: *Walden*'s cycle from winter round the year to spring, phenological charts tracking phenomena through the measure of the months, the Journal's query into exactly when the dial turns from one season to the next, exactly how to calibrate the turn of the cosmic year against each present instant. Yet the lesson revealed by his life's work of commensuration was the incommensurability of life: no human being can observe for 1,000 years, collate the infinite measures coming in from all over the globe, re-organize science from a bounded universe to an unbounded cosmos.

Surprisingly, it was Thoreau's careful protocols of measurement that uncovered this problem: while he was quite certain that Walden had a bottom, exactly where was it? As Maurice Lee reminds us, professional surveying involves the surveyor not in exactitude but in the calculation of error, figuring precision within a probabilistic range of values averaged from masses of data. Thoreau's advertising handbill flaunted his precision, but in the fine print he added a professional qualifier: "Areas warranted accurate within *almost* any degree of exactness" (Lee 2012: 133).[2] Even as *Walden* triumphantly declares that the pond's "greatest depth was exactly one hundred and two feet," its author hedges: from year to year the pond rises and falls, requiring adjustments.

Worse, he notices that the holes he has drilled in the ice allow meltwater to drain through, and, "as the water ran in, it raised and floated the ice. This was somewhat like cutting a hole in the bottom of a ship to let the water out" (Thoreau 1971: 287, 293). That is, the very act of measurement changed what he was measuring: Walden will not, ever, be measured with absolute certainty. Not only does Thoreau know this, but he saves it for his punchline, closing with the old joke about the traveler who starts across a swamp after being assured by a nearby boy that it has a perfectly sound bottom. After his horse sinks in to the girths, "he observed to the boy, 'I thought you said that this bog had a hard bottom.' 'So it has,' answered the latter, 'but you have not got half way to it yet'" (ibid.: 330).

Thoreau's response to Zeno's paradoxical universe was not skepticism but wonder. Late in life he remarked:

> All science is only a makeshift, a means to an end which is never attained. After all, the truest description, and that by which another living man can most readily recognize a flower, is the unmeasured and eloquent one which the sight of it inspires. No scientific description will supply the want of this, though you should count and measure and analyze every atom that seems to compose it.
>
> (Thoreau 1906: 105)

The scientific description is, he adds, more like a passport photo: the man of science checks for correspondence, stamps the passport, and lets it go. But in foreign ports, one's real friends "do not ask to see nor think of its passport:" the closer the relationship, the less we need "the measured or scientific account" (Thoreau 1906: 117, 119). That said, Thoreau never traveled without that tape-measure in his pocket, for the key to his poetics of measurement, from first to last, was his faith that the "measured or scientific account" is nothing less than the "passport" across borders and boundaries, the passage to "see new worlds at once." By means of that passage, once there, even the oldest worlds become new again.

Notes

1 Space prohibits a bibliography here. Interested readers should start with the work of Patrick Chura (2010), William Rossi, John Hessler, Kristen Case, Richard Primack, and Robert Sayre.
2 For Thoreau's handbill see Chura (2010: 85).

References

Barad, K. 2003. Posthumanist Performativity: Toward an Understanding of How Matter Comes to Matter. *Signs* 28:*3*: 801–31.

Beer. G. 1983. *Darwin's Plots: Evolutionary Narrative in Darwin, George Eliot and Nineteenth-Century Fiction.* London: Routledge.

Cannon, S. F. 1978. *Science in Culture: The Early Victorian Period.* Folkestone: Dawson and New York: Science History Publications.

Cavell, S. 1992. *The Senses of Walden: An Expanded Edition.* Chicago, IL: University of Chicago Press.

Chura, P. 2010. *Thoreau the Land Surveyor.* Gainesville, FL: University Press of Florida.

Emerson, R. W. 2008. Thoreau. In: *Walden, Civil Disobedience, and Other Writings,* by H. D. Thoreau, edited by W. Rossi. 394–409. New York: Norton.

Emerson, R. W. 1983. *Essays and Lectures.* New York: Library of America.

Latour, B. 1999. *Pandora's Hope: Essays on the Reality of Science Studies.* Cambridge, MA: Harvard University Press.

Latour, B. 1993. *We Have Never Been Modern.* Cambridge, MA: Harvard University Press.

Lee, M. 2012. *Uncertain Chances: Science, Skepticism, and Belief in Nineteenth-Century American Literature.* Oxford: Oxford University Press.

Rheinberger, H.-J. 2010. *An Epistemology of the Concrete: Twentieth-Century Histories of Life.* Durham, NC: Duke University Press.

Taylor, C. 1997. *A Secular Age.* Cambridge, MA: Harvard University Press.

Thoreau, H. D. 2007. *Excursions.* Princeton, NJ: Princeton University Press.

Thoreau, H. D. 1981. *Journal,* 8 vols to date. Princeton, NJ: Princeton University Press.

Thoreau, H. D. 1980. *A Week on the Concord and Merrimack Rivers.* Princeton, NJ: Princeton University Press.

Thoreau, H. D. 1972. *The Maine Woods.* Princeton, NJ: Princeton University Press.

Thoreau, H. D. 1971. *Walden.* Princeton, NJ: Princeton University Press.

Thoreau, H. D. 1906. *The Journal of Henry David Thoreau,* 14 vols. Boston, MA: Houghton Mifflin.

Walls, L. D. 2009. *Passage to Cosmos: Alexander von Humboldt and the Shaping of America.* Chicago, IL: University of Chicago Press.

Walls, L. D. 1995. *Seeing New Worlds: Henry David Thoreau and Nineteenth-Century Natural Science.* Madison, WI: University of Wisconsin Press.

3 Operationalism: Old Lessons and New Challenges

Hasok Chang

Why Operationalism, Again?

Of all the authors who have written about measurement through the course of the 20th century, I think one whose work has the most lasting value is Percy Bridgman, the Harvard experimental physicist whose philosophical works gave rise to the doctrine of operationalism. I know that this judgment is quite contrary to the going opinion. Even those philosophers and scientists who do remember something about Bridgman will mostly think of him as someone who proposed a naïve idea of operationalism, whose superficial attractions gathered brief attention that soon faded away. I want to bring back operationalism, or rather the best of Bridgman's ideas, in a new key. The main motivation for doing so, in the context of the philosophical study of measurement, is to restore attention to the semantic role of measurement. But I do not want to do this in the manner of the simplistic Bridgman who declared that the meaning of a concept consisted solely in the method of its measurement. He actually had a much subtler approach, which I want to develop further on the basis of clues collected from various parts of his work. The revised and revitalized operationalism that I offer in this contribution will be a crucial link in a new empiricist realism that I have proposed in more detail elsewhere (Chang 2012: ch. 4).

Turning my thoughts back to Bridgman, my mind jumps over to a joke dimly remembered from my undergraduate days at Caltech. A group of students were each given a barometer and asked to use it to measure the height of Millikan—not the famous physicist Robert Millikan, the first President of the college, but the central library building on campus named after him. One student did what one might expect: he took it to the top of the building and took the pressure reading, and looked up the pressure–height correlation. One student tied a very long string to the barometer, took it to the top of the building, and lowered it to the ground and swung it; from the period of this pendulum motion, the length of the string (therefore the height of the building) could be deduced. Another student said: if you're going to do that, why bother with the pendulum? Just measure the length of that string from which the barometer hangs, and that is the height of the building. Another student insisted that nothing except the barometer should be used in this measurement; he dropped the barometer from the top of the building, measured the time it took to fall, and deduced the height from Galileo's law of free fall. Another objected to the use of theory in that last procedure; he measured the length of the barometer, and used it as a ruler, climbing down the side of the building. Meanwhile, another student went over to the custodian of the building and said, "I'll give you this nice barometer if you tell me exactly how tall this building is." (You can bet that the last student delivered the most accurate result.)

The immediate lesson from this story is that there are many ways of measuring the same thing, some of them quite unexpected—the unexpectedness is what induces the laughter. In some people's imagination, the fact that there are various measurement methods applying to the same concept is the killer argument against operationalism. But of course Bridgman knew about multiple measurement methods. In fact, it is with the discussion of this issue, through the case of length, that he opened his classic operationalist manifesto, *The Logic of Modern Physics* (1927).

Bridgman was both fascinated and troubled by the fact that "essential physical limitations" forced scientists to use different measurement operations for the same concept in different domains of phenomena. Length is measured with a ruler only when we are dealing with dimensions comparable to our human bodies, and with objects moving slowly relative to the measurer. To measure the distance to the moon, for example, we need to infer it from the amount of time that light takes to travel there and return, and that is also the procedure taken up in special relativity; "the space of astronomy is not a physical space of meter sticks, but is a space of light waves" (Bridgman 1927: 67). For even larger distances we use the unit of "light-year," but we cannot actually use the operation of sending off a light beam to a distant speck of light in the sky and waiting for years until (hopefully) a reflected signal comes back to us (or our descendants). Much more complex reasoning and operations are required for measuring such distances: "To say that a certain star is 10^5 light years distant is actually and conceptually an entire[ly] different *kind* of thing from saying that a certain goal post is 100 meters distant" (ibid.: 17–8, original emphasis).

When there are many methods of measuring a given property, how do we know that they *are* actually measuring the same thing? This was a question that Bridgman struggled with in his day-to-day work. His chief scientific contribution, which was in high-pressure physics, was made possible by technical prowess: in his laboratory Bridgman created pressures nearly 100 times higher than anyone else had achieved before him (reaching over 100,000 atmospheres), and investigated the novel behavior of various materials under such high pressures; this work brought him the Nobel Prize in physics in 1946. But Bridgman was placed in a predicament by his own achievements: at such extreme pressures, all previously known pressure gauges broke down (see Kemble, Birch and Holton 1970: 457–61). As he kept breaking his own pressure records, Bridgman had to establish a succession of new pressure measures. But did these methods all really measure pressure? Or did they measure some other quantity correlated with pressure in some unknown way?

These are questions that any self-aware scientist might ask. But one of Bridgman's distinctive contributions was to raise a more fundamental challenge to the standard realist picture of measurement implicit in such questions. When we ask "Is this method really measuring pressure?", we presume that there is a well-defined property called pressure "out there," which the different methods may or may not get at correctly. Bridgman's insights shattered this semantic complacency, by pointing out that the meaning of the concept was bound to its method of measurement. If the concept is at least partly defined by its measurement method, then the question about the correctness of the measurement method threatens to become vacuous. In a statement that he would come to regret and retract later, Bridgman declared: "we mean by any concept nothing more than a set of operations; the concept is synonymous with the corresponding set of operations" (Bridgman 1927: 5). This is a drastic and unsupportable statement, if "operation" only means measurement operation. (I will later discuss a broader sense of "operation" present in Bridgman's work.)

Set aside this extreme formulation, which occurred at the very outset of Bridgman's philosophical writing. The main message to take from his more extended and considered discussions is caution: we should not start by presuming that different operations "measure the same thing." There may be no unified property there to be measured, or even if there is something there not all the different measurements may succeed in getting at it. Bridgman was careful in avoiding getting into a debate about what was or wasn't really "out there," and I intend to follow him in that reluctance.

The point is to think carefully about what exactly our measurement results signify. When there are multiple operations allegedly measuring the same thing, it is helpful if they can be applied to the same situations and give similar enough results, but even such numerical convergence in an area of overlap is no guarantee of conceptual identity, especially in areas beyond the overlap. So Bridgman decided: "In *principle* the operations by which length is measured should be *uniquely* specified. If we have more than one set of operations, we have more than one concept, and strictly there should be a separate name to correspond to each different set of operations" (ibid.: 10).[1]

We know that this solution has long been rejected by mainstream philosophers of science and by most practicing scientists, despite its initial appeal. Carl Hempel, for instance, gave a clear and convincing critique of Bridgman's position. It seemed to Hempel that following Bridgman's skeptical caution concerning multiple measurement methods would cause an intolerable fragmentation of science, resulting in "a proliferation of concepts of length, of temperature, and of all other scientific concepts that would not only be practically unmanageable, but theoretically endless." Bridgman's quest for epistemic safety made him neglect one of the ultimate aims of science, "namely the attainment of a simple, systematically unified account of empirical phenomena" (Hempel 1966: 94). As I have argued elsewhere, particularly in my entry on operationalism for the *Stanford Encyclopedia of Philosophy* (Chang 2009), standard discussions of operationalism such as Hempel's have missed the most interesting and significant aspects of Bridgman's thinking. Let me now come to those aspects in a more sustained way.

The Battle for Meaningfulness

In my view, the first step in a productive re-framing of Bridgman's message is to take him at his word when he says he was not trying to create a philosophical theory of meaning; he thought he had created a "Frankenstein" in operationalism, a monster that he did not even recognize (Bridgman in Frank 1956).[2] But he did talk and worry incessantly about meaning. What *was* he doing, then? The drive behind Bridgman's operationalism was not so much the desire for a theory of meaning, but a striving toward *meaningfulness*, driven by a fear of the loss of meaning. In the first instance, we can take operationalism as a commitment to maintain and increase the empirical content of theories by the use of operationally grounded concepts.

What do I mean by "meaningfulness" here? Before stating that, it is necessary to set the scene by clarifying that I am starting with a broadly late-Wittgensteinian theory of "meaning as use" (Wittgenstein 1953): the meaning of a concept is rooted in all of its uses that are approved by the relevant community—including, but not only, measurement operations. In fact, in Bridgman's considered view, the concept of "operation" covered all kinds of well-specified doings, including not only doing things with physical apparatus ("instrumental" operations) but "paper-and-pencil" and "mental/verbal" operations, too. He considered it a widespread misconception about his view that it demanded locating meaning only in instrumental operations (see Bridgman 1938: 122–4).[3] Ten years after *The Logic of Modern Physics*, Bridgman in fact gave a new and rather late-Wittgensteinian gloss on his view on meaning: "To know the meaning of a term used by me it is evident, I think, that I must know the conditions under which I would use the term" (ibid.: 116). It is instructive to examine the following passage from *Logic*:

> What is the possible meaning of the statement that the diameter of an electron is 10^{-13} cm? Again the only answer is found by examining the *operations* by which the number 10^{-13} was obtained. This number came by *solving certain equations* derived from the field equations of electrodynamics, into which certain numerical data obtained by experiment had been substituted. *The concept of length has therefore now been so modified as to include that theory of electricity* embodied in the field equations, and, most important, assumes the correctness of extending these

equations from the dimensions in which they may be verified experimentally into a region in which their correctness is one of the most important and problematical of present-day questions in physics.

(Bridgman 1927: 21–2, emphases added)

The beginning of this passage sounds like Bridgman's usual complaint about the meaninglessness of concepts without measurement operations attached to them. But the second sentence should give one a pause: in saying that we can trace the *operations* by which the 10^{-13} figures was obtained, he is putting into practice the notion of paper-and-pencil operations. And note that he does *not* say that this estimate of length is meaningless; rather, he says that the concept of length has now been *modified*, by its use in this paper-and-pencil operation. That sounds to me like a perfectly good meaning-as-use way of understanding the situation. Here Bridgman is getting at the heterogeneity, complexity, changeability, and contingency of meaning. Managing this contingent complexity while seeking progress in science is the task of the more sophisticated operationalism that I think Bridgman was developing.

Instrumental measurement operations always need to be placed in the larger context of the assortment of various operations which give to each concept its full meaning. Contrary to common perceptions, Bridgman was well aware of this broader operational context of measurement. So we cannot say straightforwardly that Bridgman's operationalism implied giving privileged epistemic status to measurement methods. We may choose to use instrumental measurement operations to define a concept, thereby privileging them over other kinds of operations, but even then we must keep in mind that a definition is only a criterion by which we *regulate* the uses of a concept, not the expression of the *whole* meaning of the concept.

It is possible that Bridgman was not entirely self-consistent in his various philosophical statements—but then again, who is? I am trying to develop the best possible position on the basis of what he did say. It is important to remember that Bridgman offered operational analysis as a tool of self-diagnosis and self-improvement in scientific practice. He wrote his philosophy books during his summer vacations in rural New Hampshire (much like Gustav Mahler writing his symphonies at the lakeside during summer breaks from conducting the Vienna Opera). That is not a random piece of trivia, but a reminder that Bridgman was a busy practicing scientist. He did express worries that verged on philosophical skepticism, but they were not the worries of an armchair philosopher. He was the kind of person who could say: "It is easy, if all precautions are observed, to drill a hole … 17 inches long, in from 7 to 8 hours"—that is, a hole as narrow as the lead in a pencil, in a block of hard steel.[4] Bridgman was too busy to worry seriously about the meaningfulness of concepts in well-established domains of phenomena; he was not gripped by the Humean variety of the problem of induction. He did get extremely nervous, however, about the loss of meaningfulness when a concept was extended to new domains (see Chang 2004: ch. 3).

When concepts were extended to unfamiliar territories, Bridgman could feel meaning slipping right through his fingers, as it were, if he did not succeed in tightening his grip. He realized that this was happening almost everywhere, as he performed his critical survey of almost all basic concepts in physics. The very heart of classical physics already seemed an exercise in carelessness, accepting Newton's "absolute" space and time concepts where no operations corresponding to them existed. For example, it was a grave mistake to think that it was meaningful to speak of the simultaneity of events that are separated in space, without specifying an operation, not even a mental one, by which such a thing was determined.[5] It was Einstein's genius to see the gap here, and to fill it by specifying the light-signal convention; all else in special relativity followed from that operational step. But Bridgman's point was most emphatically *not* that we should encourage future scientists to do Einstein's kind of revolutionary work. No, the point was to do science carefully in the first place so that we can "render unnecessary the services of the unborn Einsteins" (Bridgman 1927: 24).

Bridgman despaired of our "verbal and mathematical machinery," which had no built-in cut-off that would make us stop and think (and perhaps shut up) when concepts were being applied to situations in which we had no ability to give them empirical grounding. "For instance, the equations of motion make no distinction between the motion of a star into our galaxy from external space, and the motion of an electron about the nucleus, although physically the meaning in terms of operations of the quantities in the equations is entirely different in the two cases. The structure of our mathematics is such that we are almost forced, whether we want to or not, to talk about the inside of an electron, although physically we cannot assign any meaning to such statements" (ibid.: 63).

And what did modern physics teach us about the places where we had never been, like the inside of the electron? This is what Bridgman had to say, in a popular article: "if we sufficiently extend our range we shall find that nature is intrinsically and in its elements neither understandable nor subject to law" (Bridgman 1929: 444). Elsewhere, he professed his belief that "the external world of objects and happenings is … so complex that all aspects of it can never be reproduced by any verbal structure" (Bridgman in Frank 1956: 78); therefore, "all sorts of phenomena cannot at the same time be treated simply" (Bridgman 1927: 100). He lamented: "Even in physics this is not sufficiently appreciated, as is shown, for example, by the reification of energy. The totality of situations covered by various aspects of the energy concept is too complex to be reproduced by any simple verbal device" (Bridgman in Frank 1956: 78). All of that might well have come out of Nancy Cartwright's or John Dupré's mouth (see Cartwright 1999; Dupré 1993).

But was Bridgman all gloom and doom, a bundle of worries and warnings, his Nobel Prize-winning finger shaking in the face of lesser and more careless mortals? Did he have anything positive to offer, beyond his grim resolve in the face of insurmountable uncertainty, expressed in his maxim that "The scientific method, as far as it is a method, is nothing more than doing one's damnedest with one's mind, no holds barred" (Bridgman 1955: 535)?[6] I admit that the constructive side of his work was not as well developed as the critical side. This is where I think we can go beyond Bridgman, building on some of my own recent work.

Before engaging in that development, however, I need to clarify two points further. The first is what "meaningfulness" means when I say Bridgman's philosophy was a quest for meaningfulness. In the meaning-as-use framework, the meaning of a concept simply consists in all the ways in which it is legitimately used. So, any concept that is legitimately used at all has *a* meaning. But *meaningfulness* comes in degrees and in different shapes. A concept may have a well-established set of uses, but we can add more uses, and thereby extend that concept. But that is easier said than done. When we introduce new uses, we need to be careful to ensure that they are coherent with existing uses. In other words, meaningfulness has (at least) two distinct dimensions: the variety of different uses, and the degree to which the different uses cohere with each other.[7]

Second, we have to settle the question about the place of measurement in this generalized operationalist philosophy. If all kinds of uses of a concept can help make it more meaningful, then why is it so important to include measurement operations among its uses? Operationalism as I am re-packaging it here should welcome the coherent addition of any kinds of uses; in Bridgman's terms, instrumental, paper-and-pencil and mental operations are all welcome as ways of increasing the meaningfulness of concepts. So why focus on measurement operations? Why not satisfy ourselves with inventing more mental and paper-and-pencil operations of theoretical, mathematical or even metaphysical varieties? One answer is simple, and not very interesting: if we want to maintain a commitment to a very basic kind of empiricism, then concepts need to be linked up with observations, and precise quantitative measurements are the best kinds of observations we can have in science. Another answer is more specific to the scientific situation that Bridgman found himself in. Bridgman was worried about the absence of instrumental meaning because that is what he saw most at risk of being lost. This makes sense of his appreciation of the Einstein of special relativity, who chided Newton for not providing any instrumental meaning to absolute time and space, and the betrayal he felt when Einstein later satisfied himself with only mathematical operations concerning key concepts in general relativity (see Bridgman 1949: 333–54).

One manifestation of operationalism, as reconstituted here, is a demand for more well-rounded meaningfulness. If one found concepts that have only mental and paper-and-pencil operations attached to them, then the right operationalist thing to do would be to demand that more instrumental operations be found or created. On the other hand, if there were concepts that only had instrumental meaning, then the operationalist should demand that more theoretical uses be developed. It is my historical contention that Bridgman found himself in the former type of situation much more than the latter, in physics and other sciences and everyday life; this explains why he focused on the demand for more and better measurement operations. But if he implied that concepts that could not be physically measured were meaningless, then he was wrong. What he did say was that a statement was meaningful if we had an operational way (mentally, instrumentally, or paper-and-pencilly) to determine whether it was correct or not.

At one place in *Logic* he gave a list of meaningless questions, which began with "Was there ever a time when matter did not exist?" and ended with: "Can we be sure that our logical processes are valid?" (Bridgman 1927: 30–1). I think Bridgman would have said that even a metaphysical statement can be operationally meaningful, if the metaphysical thinking is done carefully so that there are definite mental operations to determine its validity; in that case, however, what we are doing would end up looking more like logic than metaphysics as it is usually practiced.

Operationalism in the Service of Active Realism

A revival of operationalism is at the heart of a philosophical position that I have advanced elsewhere (see Chang 2012: ch. 4), which I call "active scientific realism," or "active realism" for short.[8] Active realism is not only a particular position concerning scientific realism but an attempt to reframe the whole debate, taking it away from its focus on truth. I follow Bas van Fraassen in defining scientific realism in terms of the aims of science,[9] but I do not think that the realist aim of science is, or should be, truth. The kind of truth ("with a capital T") that realist philosophers typically want is not an operable aim, and therefore it cannot guide actual scientific practice.

This ought to be a simple point, although it is by no means widely acknowledged. Truth, in the standard conception of realists, comes down to a correspondence between what our statements say and how the world is. But what are the methods by which we can judge whether this correspondence obtains in each situation? At least the burden of argument should be on those who claim or assume that there are such methods, since there aren't any *obvious* ones. The standard realist strategy is to get at truth indirectly; we can pursue truth via other theoretical virtues, if they are truth-conducive. But here we are locked in a vicious circle: if we are not able to judge whether we have truth in each situation, how will we be able to tell which methods have a tendency to lead us to truth? Whether this circularity is inescapable is the main point of contention in the scientific realism debate. I do not propose to tackle that question—not here, or perhaps not ever. I want to get away from that question.

In the kind of reframing that I am attempting, it is sometimes useful to try taking things very literally for a moment. In this case, what is "real-ism?" A proper, full-blooded "ism" is not simply a description, but an *ideology* in a broad sense of the term—a commitment to think, act, and live in a certain manner. So I take realism as a commitment to engage with, and learn from reality. But what is reality? Instead of entering into serious metaphysics, I give an *operational* definition of reality. I propose to think of (external) reality as whatever it is that is not subject to one's own will. As pragmatists have pointed out, nature's *resistance* to our ill-conceived plans is one of the most important sources of our very notion of reality. As William James stated: "The only objective criterion of reality is coerciveness, in the long run, over thought."[10] But it is not only when our expectations are frustrated that we make contact with reality. Something that we cannot control can turn out the way we expect, and that is precisely what happens when we make successful predictions. Knowledge is a state of our being in which we are able to engage in successful epistemic activities, which can only happen if there is insufficient resistance from reality. As Wittgenstein once put it: "it is always by favour [*Gnaden*] of Nature that one knows something" (Wittgenstein 1969: 66e: § 505).

What this notion of reality refers to is the fundamental distinction between the self and the world. If everything behaved simply as I willed it, I would not have a sense of external reality. But does that mean that my body, because I can move it around as I wish, is not part of reality? Far from it. When G. E. Moore famously held up his hand and declared it an external object, he would have noted that the color, shape, temperature, and all other properties of the hand are there as they are, out of his own control, except for its motion. My (usual) ability to move the hand about as I wish makes it "my" hand; the other properties (including the rather fixed range of its motion) make it an external object, a part of reality. And my own *will* is part of everyone else's external reality, even though it cannot be said to be part of my own external reality.

This kind of blend of self and world is the mysterious and wonderful thing that allows us to be in the world and interact with it. This was also the world of experimental physics that Bridgman inhabited: "The process that I want to call scientific is a process that involves the continual apprehension of meaning, the constant appraisal of significance, accompanied by a running act of checking to be sure that I am doing what I want to do, and of judging correctness or incorrectness" (Bridgman 1955: 56). And he added, controversially: "This checking and judging and accepting that together constitute understanding are done by me, and can be done for me by no one else. They are as private as my toothache, and without them science is dead." One remaining challenge for me is to understand how Bridgman's toothache exists in the world of social action, but that is a story for another day.

To have contact with reality, we need to place ourselves in situations where things that we cannot control will happen; that is not difficult, indeed difficult to avoid. For *learning* to take place, we need to arrange such situations as to expose our senses to the happenings—not only the five senses, but any and all the modalities we have through which information is registered, including muscular tension, so integral to any bodily interactions we have with reality. And in order to *maximize* our learning, we should arrange situations in which our expectations are most likely be contradicted. Karl Popper's injunction for scientists to seek higher falsifiability and more severe tests, Imre Lakatos's demand for progressiveness in scientific research programs, and Ian Hacking's "experimental realism" encouraging active interventions[11] can all be seen as a demand for more contact with reality.

The core idea of active realism is certainly not my invention, although the terminology is. Bridgman has indeed been a most important influence on my thinking: his desire to specify a method of measurement to every physical concept can be taken as a manifestation of an active-realist commitment intended to turn every theoretical statement into a site of contact with reality. There is a little-known passage that provides a key to the active-realist reading of Bridgman, from as far back as 1927:

> The essential point is that our [theoretical] constructs fall into two classes: those to which no physical operations correspond other than those which enter the definition of the construct, and those which admit of other operations, or which could be defined in several alternative ways in terms of physically distinct operations. This difference in the character of constructs may be expected to correspond to essential physical differences, and these physical differences are much too likely to be overlooked in the thinking of physicists.
>
> (Bridgman, 1927: 59–60)

Concerning constructs, "of which physics is full," Bridgman not only admitted that one concept could correspond to many different physical operations, but even suggested that such multiplicity of physical-operational meaning was "what we mean by the reality of things not given directly by experience." For example, Bridgman argued that the concept of stress within a solid body had physical reality, but the concept of electric field did not, since an electric field only ever manifested itself through force and electric charge, by which it was defined (see ibid.: 57), while internal stress had various physical-operational manifestations in terms of quantities that are measurable on the surface of the solid. (I think he was wrong about the electric field, but the general point is clear.)

Does this come down to the stance that a theoretical concept without direct empirical meaning is worthwhile only if it serves to make connections between empirically meaningful concepts? Yes, at first glance, and that would have been how the logical positivists wanted to read Bridgman. But we can also give a broader reading of what Bridgman is saying. The active realist stance does privilege physical operations as the most obvious ways in which we can make contact with reality and learn about it (as long as we reject the possibility of a direct mental apprehension of reality). But there is room for more sophistication here. According to active realism, it is an entire system of practice that makes "contact" with reality. Having the right sort of paper-and-pencil and mental operations, working together with the instrumental ones, is required for learning about reality. This is why we want concepts that are highly meaningful in a well-rounded way, because such concepts facilitate our learning about reality. Without such many-faceted concepts, we could not make statements that link up theory and observation and thereby create expectations that some specific observable thing will happen if the theory were true.

Multiplicity and Unity

Having re-conceived the significance of operationalism in the context of active realism, let me return to the problem of multiple measurement methods, with which I began this contribution. I think we should learn to celebrate multiplicity as richness of meaning, overcoming Bridgman's own timid conscience tending toward a fragmentation of the concept into many different ones. But the richness needs to be shaped up in a coherent way, and that is where Bridgman's conscience can play a positive role. He was right to stress that we should not simply assume that different measurement methods that are said to measure the same thing actually do measure the same thing. But such connections can be established, at least in some cases, and Bridgman's pessimism overcome.

I will not enter here into the details of how connections between different measurement methods can be made. Possible strategies would range from ascertaining numerical agreement in measurement outcomes (which Bridgman considered) to making a highly theoretical demonstration that each method does get at the quantity in question. More generally speaking, the desideratum is to establish as many credible links as possible in the whole system of practice involving the concept in question. This involves establishing not only links between the different measurement methods, but also between the measurements and the other uses of the concept.

If we listen carefully to Bridgman, there is actually no conflict between him and Hempel. Hempel missed out on Bridgman's demand for diverse and coherent operationalization, as he was preoccupied with disputing a different part of Bridgman's thinking, namely the impulse to demand that *every* concept should be physically measurable. But that was only an impulse, and Bridgman also said: "there need be no qualms that the operational point of view will ever place the slightest restriction on the freedom of the theoretical physicist to explore the consequences of any free mental construction" (Bridgman in Frank 1956: 79). Hempel's own philosophy, in stressing the importance of the variety of evidence, was in line with Bridgman's forgotten demand for multiple methods.

Bridgman's reflections on the meaning of concepts help provide us with a practical and realistic framework for scientific progress, steering between the presumption of unity and the despair about multiplicity. If we get Bridgman on a more optimistic day, multiplicity is recognized as a fact of life but unity, or rather coherence, is recognized as an ideal toward which science can strive. Concepts become more meaningful if they exist within a more thickly connected set of practices (i.e., uses of concepts); more meaningful concepts are more effective facilitators of inquiry about nature. Operationalism as I conceive it is the doctrine that every concept should be made as meaningful as possible, by the crafting of more uses of it in harmony with each other. Operationalism in this sense would not destroy systematic unity; on the contrary, it is an optimal strategy for achieving as much systematic unity as nature would allow, in a strongly empiricist system of knowledge.

This approach also works for extending concepts to fresh new domains in which theories are uncertain and experience scant. We start with a concept with a secure network of uses that gives it stable meaning in a certain domain of circumstances. The extension of such a concept consists in creating various well-specified uses of the concept in an adjacent domain which are credibly linked to each other and to the pre-existing uses. At each step along the way we can check the entire set of uses, the whole meaning of the concept, for overall coherence. Contrast such a deliberate process with the vague presumption that terms in a theoretical equation must have the same meaning in the entire mathematical range of the variables.

I will end with a self-conscious note. My doctrine of "active realism" was developed hand-in-hand with a strong advocacy of pluralism (see Chang 2012: ch. 5). But it would seem that in the last remarks I have made here I am advocating unity, rather than plurality. But the contradiction or tension here is only apparent. My pluralism consists in advocating the cultivation of multiple systems of practice. But within each system it makes sense to seek the maximal degree of unity or coherence. Operationalism can aid that task of making a coherent development of each system, and it is conducive to the cultivation of additional systems, too.

Acknowledgments

I would like to thank Alfred Nordmann, Oliver Schlaudt, and their colleagues for inviting me to present at the "Dimensions of Measurement" conference. Among them, I thank Nicola Mößner also for her careful attention to an earlier draft of this contribution. I also thank various participants at that conference for helpful comments, and similarly also members of the Philosophy and History of Physics Reading Group in Cambridge. Whenever I write about operationalism, I thank Gerald Holton for all he taught me about Bridgman (and everything else).

Notes

1 That would certainly be one way to avoid unwarranted identifications, but it is not the full picture that Bridgman gave.

2 Chapter 2 of the volume P. G. Frank (ed.), *The Validation of Scientific Theories* (Boston: Beacon Press, 1956, reprinted in 1961 by Collier Books, New York) contains papers arising from the symposium on "The Present State of Operationalism" at the annual meeting of the American Association for the Advancement of Science in Boston in December 1953, co-sponsored by the Institute for the Unity of Science and the Philosophy of Science Association.

3 See Bridgman (1959a: 3) and Bridgman (1959b: 522).

4 Quoted in Holton (1995: 222–3)

5 Newtonian physics had an operationally well-defined concept of local simultaneity, but extended it to events separated from each other in space without doing the necessary operational work.

6 This collection of essays was originally published in 1950; the 1955 edition includes some additional papers.

7 I need to make a thorough account of what "coherence" means. For an initial attempt see Chang 2012: 15–7 and other places referred to in the index.

8 The summary of active realism I give here may sound very obvious and trivial, in which case I have achieved my objective in articulating it.

9 "Science aims to give us, in its theories, a literally true story of what the world is like; and acceptance of a scientific theory involves the belief that it is true. This is the correct statement of scientific realism" (van Fraassen 1980: 8).

10 Quoted in Putnam (1995: 11); originally in James (1978: 21).

11 "Don't just peer; interfere!" (Hacking 1983: 189).

References

Bridgman, P. W. 1959a. *The Way Things Are*. Cambridge, Mass.: Harvard University Press.

Bridgman, P. W. 1959b. P. W. Bridgman's "The Logic of Modern Physics" after Thirty Years. *Daedalus* 88: 518–26.

Bridgman, P. W. 1955. *Reflections of a Physicist*. New York: Philosophical Library.

Bridgman, P. W. 1949. Einstein's Theories and the Operational Point of View. In: *Albert Einstein: Philosopher-Scientist*, edited by P. A. Schilpp. 333–54, La Salle, Illinois: Open Court. Also reprinted in Bridgman 1955: 309–37.

Bridgman, P. W. 1938. Operational Analysis. *Philosophy of Science* 5: 114–31. Also reprinted in Bridgman 1955: 309–37.

Bridgman, P. W. 1929. The New Vision of Science. *Harper's* 158: 443–54. Also reprinted in Bridgman 1955: 81–103.

Bridgman, P. W. 1927. *The Logic of Modern Physics*. New York: Macmillan.

Cartwright, N. 1999. *The Dappled World: A Study of the Boundaries of Science*. Cambridge: Cambridge University Press.

Chang, H. 2012. *Is Water H_2O? Evidence, Realism and Pluralism*. Dordrecht: Springer.

Chang, H. 2009 Operationalism. In: *The Stanford Encyclopedia of Philosophy*, edited by E. N. Zalta, http://plato.stanford.edu/archives/fall2009/entries/operationalism/.

Chang, H. 2004. *Inventing Temperature: Measurement and Scientific Progress*. New York: Oxford University Press.

Dupré, J. 1993. *The Disorder of Things: Metaphysical Foundations of the Disunity of Science*. Cambridge, Mass.: Harvard University Press.

Frank, P. G. (ed.) 1956. *The Validation of Scientific Theories*. Boston: Beacon Press; reprinted in 1961 by Collier Books, New York.

Hacking, I. 1983. *Representing and Intervening*. Cambridge: Cambridge University Press.

Hempel, C. G. 1966. *Philosophy of Natural Science*. Englewood Cliffs, N.J.: Prentice-Hall.

Holton, G. 1995. Percy W. Bridgman, Physicist and Philosopher. In: G. Holton, *Einstein, History, and Other Passions*, 221–7, Woodbury, N.Y.: American Institute of Physics Press.

James, W. 1978. Remarks on Spencer's Definition of Mind as Correspondence. In: W. James, *Essays in Philosophy*, 7–22, Cambridge, MA and London: Harvard University Press.

Kemble, E. C., Birch, F. and Holton, G. 1970. Bridgman, Percy Williams. *The Dictionary of Scientific Biography* 2: 457–61.

Putnam, H. 1995. *Pragmatism: An Open Question*. Oxford: Blackwell.

Van Fraassen, B. 1980. *The Scientific Image*. Oxford: Clarendon Press.

Wittgenstein, L. 1969. *On Certainty*. New York: Harper & Row.

Wittgenstein, L. 1953. *Philosophical Investigations*. trans. by G. E. M. Anscombe, New York: Macmillan.

Part II

Images as Measurements

4 Photo Mensura

Patrick Maynard

The relativistic *homo mensura* of Protagoras[1] seems to have two general meanings for current discussions of measurement: first, for the collection of data and, second, for its presentation—in other words, relativities regarding what *shows up* when we measure, and regarding how we *show* (display) our measurements to ourselves. Photography, as an important instrument of modern scientific measurement, provides clear and interesting illustrations of this, especially when we consider it historically, also according to its most common functions.

History proves the best way to frame not only our topic of photographic measurement but photography itself. Framed in its most usual, unhistorical, manner, within the vast history of picture-making, photography is almost universally misunderstood in terms of certain of its products: pictures of things. But although we hear of "the photograph," that versatile, ever-developing family of technologies (ways of doing things) that we call "photography" came into the world as much for measuring as for depicting, and this function has never left it. Thus, the eminent physicist François Arago's presentation of the new technology of the daguerreotype to the French chambers in July 1839 was principally in terms of its "usefulness to the sciences"—as a means for minute, accurate, commonly-scaled data collection, storage, sharing, and processing in the sciences of archaeology, geology, geography, and biology (Arago 1839: 21). Next, and with most emphasis, Arago, then Director of *l'Observatoire de Paris*—and who had early made his name by heroic, meridian-arc measurements in the field for Earth's geoid—listed astronomy, for which daguerreotypy was a detective-measuring device, one as revolutionary as the telescope, he predicted.[2]

This conception was not particular to scientists like Arago or to formal academies. Consider the closing remarks of even a one-page, popular introduction of daguerreotypy by a contemporary journalist, Edgar Allan Poe. "Among the obvious advantages derivable from the Daguerreotype," Poe wrote, "we may mention that, by its aid, the height of inaccessible elevations may in many cases be immediately ascertained, since it will afford an absolute perspective of objects in such situations, and that the drawing of a correct lunar chart will at once be accomplished, since the rays of this luminary are found to be appreciated by the plate" (Poe 1840: 2). That is how even the general public could understand the new processes from the start: in terms of scientific measurement, not just in terms of "art," which in that age denoted making pictures of things.

"Their Relative Effects"

Arago argued that the positions and apparent magnitudes of overwhelming numbers of celestial objects, visible only at different times or from remotely separate places, could be mapped on Daguerre's plates into common spaces, with precisely clocked exposure times, to provide not just rough ordinal through ratio, but perhaps also metric, scaling of light intensities, ranging through the sun to the faintest stars. "The physicist will henceforth be able to proceed to the determination of *absolute* intensities," he remarked, adding the significant phrase: "He will compare the various lights by their relative effects."[3] Here "by their relative effects" means variations in light from celestial bodies being measured by variations in their physical effects on photographic receivers. This comment has two notable implications for measurement. Immediate is the idea of nature's automatic, or "self-," registration, newly enabled by photography. Self-registration has been discussed in recent years as constituting a peculiarly 19th century contribution to our conception of "objectivity"—but as an essentially *negative* component of that idea, it is held, whereby human intervention and interpretation are "restrained" by the mechanization of image production, so that nature "speaks for itself," without a prejudicial "personal element of interpretation" (see Daston and Galison 1992/2007).[4]

However that may be otherwise, that is not as it was presented in written records regarding photography at the time. For example, even before Arago's presentation of Daguerre's invention, and within weeks of announcement of the Englishman William Henry Talbot's alternative salted-paper method, a camera-free, clockwork mechanism had been devised by T. B. Jordan, for what he termed trace "self-registration" of the fluctuations of thermometers and barometers, by a slit-scan device on a clockwork drum, which was soon calibrated to minutes. Thus already by 1841 Robert Hunt's treatise (the first in English on photography) stated that "there are so many advantages attendant on self-registration, as to make the perfection of it a matter of much interest to every scientific inquirer" (Hunt 1854: 207), and provided detailed accounts of apparati and processes without reference to avoidance of subjectivity, indeed free of any comparisons with hand-work. (As for hand-work, we know that for the rest of that century and beyond line engraving by draughtsmen was smoothly integrated with photography for printing of scientific plates, as for other publications.[5])

Regarding self-registration—as shown in his portrait (Figure 4.1) as a wizard among his devices—Étienne-Jules Marey's famous photometric "chronographics" of animal motion were preceded by his 1863 improvement

Figure 4.1 Étienne-Jules Marey and his measuring devices.
Source: Wikimedia Commons, https://en.wikipedia.org/wiki/File:Marey_Portrait.jpg.

of the nonphotographic "sphygmograph" for measurement and record-ing of pulse and blood pressure. The discovery or design of channels for self-registration of natural phenomena in the 19th century is itself a large, interesting, topic.

Self-Calibration

Besides self-registration, a second significant connotation of Arago's "rela-tive effects"—important even more for issues of objectivity—concerns the relativity of the different registers themselves. Partly by means of photogra-phy, it was at this time that we began seriously also to measure ourselves—not only our pulses and physical actions but also our sense responses: that is, to measure them as measuring devices. As Protagoras's *homo mensura* expresses, since Classical times, relativists and skeptics have delighted in demonstrating the variations of responses, "measurements"—especially those of the senses—between individuals, the same ones at different times and in different states, different peoples, different sentient creatures. In reply to such relativism—indeed skepticism—we may say that, just as differences due to parallax—including the "conflicting" reports of binocular vision—actually expand rather than challenge our powers of objective measurement, the 19th century's inventive ingenuity with measuring devices demonstrated how, by controlled measurement of the devices, their very variations greatly expand objective understanding.

Photography provides a notable case. A simple example is the distinction it forces between what are termed *radiometry* and *optical photometry*—that is, between objective *radiance* at a point (measured in watts) and *luminance* (lumens) for human eyes. Even before photography, in the year following astronomer William Herschel's discovery of the invisible "thermic" rays of infrared, Johann Wilhelm Ritter's 1801 *actinographic* discovery of the so-called "chemical rays" of ultraviolet, by exposure of silver chloride (a direct forerunner of photography), had begun to measure the limited span of human visual sensitivity.[6] Next, as a complement to such measurement of nature and of ourselves, we should attend to measurement of the third—the photographic—party itself: that is, attend not just to radiometry and photometry, but to *sensitometry*, measurement of the sensitivity of photoreceptors to intensities and colors of light. Immediately apparent to all practitioners is the hypersensitivity of most photographic receptors to shorter wavelengths, among other characteristics peculiar to different receptors. Timed use of photosensitive materials for even the astronomical tasks Arago outlined would soon have made evident their partialities, not only ours, as was shown by the standard Hurter-Driffield actinometric "characteristic curves" for plate or film speeds, later in that century (Figure 4.2), briefly described below.[7]

With this third measure, I suggest, we find a significant kind of objectivity overlooked in accounts of 19th century objectivity as seeking "to extirpate human intervention between [nature] and representation," even in a "near-fanatical effort to create … images that were certified free of human interference" (Daston and Galison 1992: 81, 96). Not only did we gain from photography and similar mechanisms an increased awareness of the relativity of our senses' responses to nature—always an important step in objectivity—in the process of inventing and fixing these artefacts, we gained recognition of the systematically *different* relativities in the responses of various measuring systems that we had engineered, as well. As "by their relative effects" gains meaning not only for variations registered by receptors but for different receptors as well, Protagoras's relativism must be qualified. It is not simply a question of *homo mensura*, for we construct multiple other information systems besides our own, which measure and extend ours, while we measure their comparative tolerances, in turn. Thereby we continue to advance in an important kind of objectivity. This is not by dispelling *relativity*—for all registrations are relative to receptors—but, in a significant sense, *subjectivism*, the sense of being confined within our responses, since we construct other perspectives on them.[8] Thus, without denying subjective experience from our human points of view—experience, which, since the time of Arago, photo artists have continued to explore—we have placed ourselves within an evolving "community" of natural and artificial information systems, the latter now enlarged by computers.

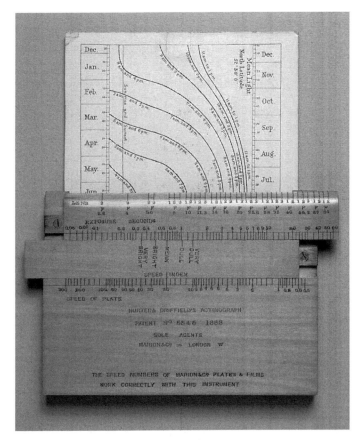

Figure 4.2 "Hurter & Driffield's Actinograph" slide-rule (photo Dick Lyon).
Source: Wikimedia Commons, https://en.wikipedia.org/wiki/File:HD_Actinograph.jpg.

Functions and Objectivity

Having begun with emphasis on the multiple functions of photography, the discussion of self-registration and objective measure has tended to run photo registration together with recording, whereas—in science as in life—sensitivity hardly entails constancy. Therefore, let us also keep in mind the distinction made at the start, between collecting data and displaying it, as we go back to the earliest days, with Arago's presentation, to distinguish several functions of photographic technologies. As noted, we need first to distinguish (i) *registering* differences—that is, information—from (ii) what Marey called *inscribing* (by *appareils inscripteurs*) them, even over distances—that is, *recording*. This marks a significant distinction for technological history generally. Marey's 1878 book title was *Méthode graphique dans les sciences experimentales*, not *Méthode mécanique*, and his posed portrait shows not only his measuring devices but the curling flourishes of their printouts. By contrast, Hermann von Helmholtz's most famous invention, the ophthalmo-scope, was, like his galvanometric clocking of nerve impulses, only of type (i), a detective-measuring device, absent recorder. For practical use the actions of photographic measurement registration require not just fixed images, but also plane surfaces of regular sizes, on which lines, writing, and other mark-ings may be applied. Such "fixing," as William Herschel's great scientist son, John, named it as he invented its first chemical process, was a recurring chal-lenge for early inventors of photo processes.

Mention of the younger Herschel reminds us that, beside the two functions of registration and recording of measurements should be placed in (iii) *repro-duction*. He was also inventor of the blueprint. Even as objectivity in meas-urement consists in applying common scales to things, it presupposes another invariance—agreement among observers, across times and conditions—for as Claude Bernard, another great scientist of the time, wrote (concerning "The Independent Character of Experimental Method"), "Art is I; Science is We."[9] Recording is necessary for that, not sufficient. Here again photography has made important contributions to objective measurement. Echoing the cultural impact of movable type, photomechanical processes have provided wide distribution of identical records for common comparison.[10]

This has great meaning for manufacture, not just discovery. Indeed, photography's first invention is accepted to be Nicephore Niépce's in the 1810s, and his initial interest was copying prints. Lithography had been around only since 1798; thus technologies for multiple copies not only of prints but also of musical scores and other autographs were of interest. So was the copying of ancient inscriptions and so forth—also argued by Arago.[11] Such line reproduction was among the multiple motives of John Herschel, a friend of Talbot's, whose independent experiments with iron salts rather than silver halides produced in 1842 his Prussian blue "cyanotype," precursor to the blueprint. Cheap, safe, fast, stable, and easy to produce on paper—hence easily transportable—as negative contact-prints of line draw-ings, such camera-free photographs (since Herschel also invented the term "photography," we follow his usage here) became for over a century the standard reproductive medium for sharing design measurements. Without it a modern world would have been impossible. Thus the photographic age is indeed the "age of mechanical reproduction," as Walter Benjamin wrote, but in a much broader sense: the mechanical production and mass reproduction of all its artefacts—notably beginning with steam engines—that define the modern world depended on this reproductive use of photo-processes.[12]

Imagining Seeing: Photo-Depiction

Still, Herschel's most important photographic invention was a byproduct of his main interest in iron-based photography: creating new pictures—indeed multi-colored pictures—of things. It seemed important to put off to last this fourth photo-function, with which, as mentioned at the outset, conceptions of photography usually begin: (iv) *depiction*—photo-*depiction*. What is that? Let us take depiction first. I would like to consider it in terms of its own main function, in whatever medium: perceptual *imagining*. But depiction is more than a stimulus to our imaging seeing things, which can also be evoked by vivid description. Treating something as a depiction is not just imagining seeing something by seeing (usually) another thing; it is also treating one's own actual seeing of the latter as that imagined seeing—therefore imagining of oneself, of one's own actions.[13]

The self-involvement adds great vivacity to the experience. By contrast, since we just considered photo-reproduction, witness the pre-digital setting of typeface from "camera ready" copy. We do not treat what we see on such pages as depictions of the original marks, nor do we try thereby to detect their originals; we simply take them as copies of their originals. Similarly, we would not these days count photocopies, including blueprints, as depictions of, "pictures of," their originals—that is not the function we give them—nor were many early photographs so treated. For example, Talbot's initial superposition images are not strong candidates for depiction: they were more like shadows, imprints, pressed leaves, which do not usually function as depictions, either—although of course they may, and this can be a matter of degree. That helps us see that depiction is not basically a matter of a kind of visual resemblance or of projection, but rather of a function we can give some things, that of vividly imagining perceiving, for which pre-existing visual resemblances are of course most helpful—although, as is sometimes noted, such resemblances typically appear only once the function is established.

In summary, the historical record indicates that, while depiction has emerged as its most common function (in 19th century terms, "application"), photography was invented, and continues to be re-invented, for other functions, as well. A society that banned depiction of anything that is in heaven above, the earth beneath, or that is in the water under the earth could be technologically modern and heavily dependent on photography. With that framing, photo-depiction may be understood as one typical industrial tool: a cheaper, easier way of manufacturing incentives to vivid imagining than traditional hand-work, if that is something we want to do with it.

These days, depiction of all kinds constitutes such an enormous fictional industry—our modern world's version of aboriginal "Dreamtime"—that we scarcely need reminding how much of it consists, in still and moving picture photography, in photographing prepared props, models, actors in order to inspire imagining seeing—seeing them not exactly (or not at all) but rather as different things. Thus what we can detect from a photo is often quite different from what we are induced to imagine seeing by its means. From that standpoint, things photographed are, like paint, tools of depiction: cheap and handy light modulators with which to construct our fictions, for which digitalization merely extends established practice. This is the opposite, for example, of Marey's intentions when he in effect invented motion pictures with his chronophotography, his purpose being the analysis of motion, not appeal to imagination—certainly not entertainment (commercial at that)—on which account, exasperated, he dismissed a close assistant.[14]

Multi-Function

A further point about multiple functions needs to be made. In considering photo-functions so far, we have first isolated four of them (leaving aside, for notable example, photoengraving, by which computer chips are manufactured), then emphasized their combinations. Something that has proved particularly interesting, also very confusing, about our fourth function is the variety of combinations it has with Arago's main function, detection, including measurement. That, biological or invented, many functional objects, arrangements, processes have more than single functions would be generally accepted. In nature, wings are for flying, also for fanning and for display; in culture, clothing protects from the elements, provides social privacy, also display.

Aside from efficiency, such multi-functionality is a profound fact of biological and artefact development, since both depend crucially upon makeshift adaptations of existing functional structures to novel uses. Yet, better to understand photo-depiction, as well as widespread confusion regarding it, we need to observe the different forms that multi-functionality can take. The history of photography itself provides many cases in which functions are, as Aristotle taught us, purely "coincidental," not only in genesis but use. One of Herschel's photo-chemicals, ferric ammonium citrate, had previous (and present) pharmaceutical use in "iron tonics." In the American Civil War one might have been shot with nitrocellulose ("gun cotton") gunpowder, liquid-bandaged with its flexible version, photographed by Scott Archer's 1851 wet-plate method, in its "collodion" form.

By contrast, we took our first three photo-technologies in related functional order, providing increasing levels of objectivity in measurement. So far our fourth, photo-depiction, might appear to stand apart, perhaps as a frivolous cousin to the others. To the contrary, depiction demonstrates how, in artefacts as in biology, even closer relations than order may hold among multiple functions, for they may also interact, and in more than one way—even reciprocally. What seems never well enough grasped is how, with photography, imagining seeing has been made one mode of access to registration forms of measurement. It is still far from the only mode of photographic access. For example, Edwin Hubble's 1917 PhD thesis opened with the statement that "the photographic plate presents a definite and permanent record beside which visual observations lose most of their significance" (Hubble 1920: 1).[15]

Arago's paradoxical promise of 1839, that "it is surely the unexpected upon which one especially must count," continues to be fulfilled, from the discovery of radioactivity and X-rays in fogged plates through the star spectroscopy that separated stellar elements, thereby redshift, thereby an expanding universe—arguably including X-ray crystallography—possibly all without depicting, getting us to imagine seeing, anything. Similarly, the Galileo orbiter of Jupiter featured, besides a seven-part atmospheric probe, an array of detectors recording "relative effects:" UV spectrometer, plasma-wave, energetic particles, and dust detectors, and finally a photographic camera, all of which registered data, stored it, sent digital flows to Earth, to be received, stored again, analyzed.[16]

Interactive Functions: Showing What Shows Up

Thus a historical "framing" of photography again poses an important, over-looked question for our recent time, even if we missed the earlier history. Given that nowadays all such data comes to us in a common, digital form, remote from human senses, why does converting just a part of it, filtered photon patterns from cameras, into pictures, appeal to our imagining seeing? The engineering answer is, so that this data can be fed into the human visual recognition system, an enormously powerful information processor naturally provided. We do not have to build this system, only train it—thus a great effi-ciency of user-friendly interface, saving time and money. To a lesser extent, that may apply to other senses, as well: "scientists sometimes translate radio signals into sound to better understand the signals," states NASA—thus what is called "data sonification."[17] It works by the same good fortune that pictorial perception does: by human visual systems being very "lossy" in certain regards. Accordingly, filtering, false-color and other enhancements are employed with the camera data, better to adapt depiction to our pecu-liar cerebral equipment—notably, via sight, an information pathway already optical, chemical and electrical by nature, and comprising one-half of our brain's sensory inputs (see Kandel 2012: 238). Special training is involved for that, as for all technical tasks—here partly because whenever imagination is invoked any activity becomes more open and other senses are recruited. As mentioned at the start, "photo mensura" applies not only to how we measure but to how we *show* measurements to ourselves. (A dog planet might require "OUI"—olfactory user interfaces—as well.) "*Homo*" *mensura*.

To be sure, a distinct motive for depiction is public support and thereby funding. These days, photo-depictions on TV, Internet, in newspapers, as appeals to imagining seeing, are essential for public support of scientific efforts, more effective than, say, reports on charged particle fields. Thus NASA maintains an extensive, impressive website gallery of color photos, designed to appeal to amateur viewers' visual imaginations.[18] (Marey's asso-ciate, George Demenÿ, had a point.) This provides a clear example of not only connections between distinct functions, but of reciprocal ones. Not only are camera registrations of light made more accessible to viewers by depic-tive display, the vivacity of the depictions—and audiences do care for vivid imagining (for example, they will pay extra for the invisible factor called "real time")—is increased by the belief that it is also detection—that the depicting states of the display, being caused by the very things depicted, are manifested by the display.

Slit-Scan Functions (1): Torsos and Tangents

To conclude with some more modern examples of reciprocal interactions of the measuring and depictive functions, let us recall our Jordan 1839 slit-scan thermo- and hygro-graphs. The most familiar use of slit-scan measurement today is likely the sports photo finish. As a slipped color-printing register reveals inks, to our vision, clear robin's-egg blue or deepest purple, actually to be varying proportions of the same cyan and magenta dots on a white paper ground, so photo finish images, smeared by racers across the moving light-receptor of modern slit-scan cameras, reveal the multiple functions that lie within what we call "photography" (Figure 4.3). For even if we begin from its figuratively depictive aspect, it is obvious (from the runners' feet) in this photograph that this function is strongly inflected by the more primary one of detective-registration: notably for measurement—and that merely at the basic ordinal level. Rather than a mechanism that registers how different 0- to 3-dimensional places are spatially related at a given time interval, this system registers different times at one "place," which is part of the 2D plane rising from a nominally 1D finish line.

The receptor for a shutterless camera, focused edge-on along that plane, is shielded from anything but that plane, by a razor-thin slit in a mask. Next, the photo-receptor is advanced at a rate adjustable to racing speeds: fast enough separately to register what things cross the plane at different times, slow enough to accrete from (nowadays) frames per second slices, sufficiently recognizable images of them for judges—although partly for depictively imagined seeing at later dates, since there is a market for that, too. For the primary function of timing, these mutually supportive yet opposed constraints well express our first and fourth photo-functions, and their shifting relationships. Here, measurement by detective-registration is primary, depiction ancillary, so these are clearly distinct photo-functions.

While on the topic of sports, examples from the 2012 Olympics may introduce some broader "photo mensura" issues. For example, the women's triathlon final in London awarded a gold medal, even though, were it a horse-race, it would have been the silver medalist by a nose. But while tangents for horses are reckoned by noses, for humans, noses do not count. Unlike bicycle and skating races, foot races are not judged by absolute first tangents—not even those of feet—but by torsos. Yet a qualifying race had revealed problems with torso-tangent tests (Figure 4.3).

That was declared tied despite the judgment of its highly experienced chief photo-finish examiner, that the torso of one runner—officially defined atop by collar-bones—reached the line just ahead of the other's. The examiner reported:

Figure 4.3 AP Photo/USA Track & Field photo finish.
Source: publicly available handout, 22/06/2012.

In the end, my read was subjective. The involvement of the torso is always subjective to some degree. [The USA Track and Field] went with what they could actually see. I was overruled … But I did my job. I called what I saw. I try to stay consistent. If I went back and read that photo a hundred times, I would call it the same way every time.[19]

(Jennings 2012)

Referees however judged the tie on the "visual evidence, on what we can actually see in the picture"—as opposed to an interpolation, "an educated guess on torso position," on which a computer projected an obscured right clavicle from data points on arm and number-bib of one racer. This reference to the "subjective" brings us back to our earlier distinction between relativity and subjectivity. Subjectivity, which entails sufficiently complex agencies to be termed "subjects," is but a kind of relativity, and most relativity is not subjective. Hurter-Driffield sensitometry curves based on densitometer measurements of amounts of silver deposited allowed nonsubjective comparison of photoreceptors, notably of plates and films. In the model shown above (Figure 4.2), relative to a given receptor (represented by a printed card), a pocket-sized actinometer employed four logarithmic scales—two sliding—for our multiple relativities of response: average light (even according to months and times of day for a given latitude), speed, aperture (by f-stop), then length of exposure.

Nonetheless, we apply objectivity even to subjective responses for which there are no such public appliances, when we accept norms, which is universal regarding human behavior. Our wide reliance on qualified and experienced judges in many matters—even those involving not human values but rather complex potentialities—demonstrates this. (In China it was said that although in every time there are many fine horses there are few fine judges of horses.) So, that a track judge's decision was "subjective" should not in itself count against its validity, if it well embodies a perceptual norm. "Footrace winner" is not a natural kind, but rather a conception embedded in ancient human traditions, while torsos—which are reasonable ways of understanding what such racing involves ("breasting the tape")—are, spatially, indefinitely bounded volumetric entities. In that context—notwithstanding "what we can actually see"—measurement technologies that far outstrip human perceptual capabilities, for comparisons traditionally based on these capabilities, strain the relevance to what they are measuring, even for the quantifiable variables of space and time.

Slit-scan Functions (2): "Light in Flight"

Besides "graphically" illustrating the often overlooked varieties of interplay of depictive and other measurement-detection photo functions, the racing photo finish makes clear another interplay, one photography has more generally brought to our attention by its alleged "instantaneous" nature: that of measurement as the comparison of two events, that measured, and the action of measurement. The apparent distortions of racing photo finish images well demonstrate how the shapes of the records of things and processes—thereby of the things and processes as they appear to our depictive perceiving—can be manipulated by adjustments of the camera mechanism, according to our wishes. For example, other most recent descendants of Jordan's slit-scan include the physicists' "streak camera" apparatus, in which, as with the photo finish, the temporal order of photons from a scene passing through a slit is spatially amplified on a receptive surface. But here, trillionth of a second laser pulses reflected by objects through the slit trigger electrons by a rapidly reversing voltage gate, which "sweeps" them spatially across a receptor according to time order, so that (after transduction back to photons) they may be recorded on a sensor at a rate of about a trillion frames per second. In this "light in flight" apparatus, it is the shapes of the stages of the reflected lights' wavefronts themselves that are directly recorded, rather than those of the objects that have scattered them.[20]

Thus, as is characteristic of slit devices, one spatial dimension recorded on a two-dimensional surface is traded for a time record. While Jordan's thermometer was essentially one-dimensional, the ingenuity of the photo finish was to build 2D presentations of 2D forms out of its 1D time slices—presentations, as we have seen, that are adjusted to our depictive purposes. "Light in flight" photography works on the same principle, but with computational combination of many of these timed lines (produced by moving the object or rotating a mirror) to recover the second space dimension.

In addition, for each line to register millions of passes must be made with very precise registration. The result, presented to our eyes as a Marey-style moving depiction of a light pulse as it interacts with a stationary object, does not allow us to see (as is sometimes said) but rather vividly to visualize (for it depicts) the advance of photons, but with the assurance that what we imagine seeing accurately corresponds to the paths of many events. As computational photography develops over the next decades, we may expect more ingenious combinations of our four photographic functions.

Conclusion: Photo Mensura

Arago remarked that "when an observer applies a new instrument in the study of nature, his expectations will be relatively small in comparison to the succession of discoveries resulting from its use" (1839: 21–2). Since most of our knowledge of the universe, at its greatest and smallest scales of time and space, comes from light and other electromagnetic sources, photography, considered as that broad, still developing family of technologies for register-ing, recording, transmitting, and making available to us such information, has emerged as among our main means for measuring the universe. Photo mensura.

The wonderful Johannes Kepler once wrote: "My witnesses are time and light, and those who hear these witnesses, as will the upstanding and the eru-dite."[21] This contribution hopes to assist witness of the latter kind.

Notes

1 Protagoras, *On Truth*: *"pánton chremáton métron ánthropos,"* as quoted by Plato, *Theaetetus* 178b. Throughout the Latin is in italic, whereas the pidgin Latin-Greek of the title is left in Roman.
2 For references (including Talbot's uncharacteristically ungracious comment) see Maynard (2005: 72–3).
3 "Metric" may be an exaggeration. Perhaps by "absolute" all Arago meant was that observers would no longer be working by visual comparisons with, literally, standard candles—although in time, thanks to photography, Cepheid variable stars and more recently type 1A supernovae, would provide standard candles—that is, reproducible points for comparisons of luminosity—by their intrinsic luminosity.
4 While Marey figures in the lead sentence of the article, and is cited and illustrated thereafter, he does not appear in the index of the 481-page book. For an excellent, original study of Marey's various photographic investigations, still and moving (see Braun 1994).
5 The modern classic study of this is Ivins (1953: 113–34).
6 Ritter, inventor of electrolysis, was friend to Helmholtz, Ørsted, Goethe, and other notable "Romantic Age" inquirers, as well as to philosophers.
7 Its inventors misnamed the boxwood apparatus with a "graph" rather than a "metric" suffix, since it did not inscribe, but rather measured, by its double-slide rule, for separate cards corresponding to different films or plates—today, different ISO numbers.

8 The conception of increasing objectivity in science through such widening frames I owe to discussions long ago with Professor Clifford Hooker, who further pointed out that succeeding theories, such as relativity, are usually able to represent preceding theories within them—rather as simple linear perspective theory wins objectivity by representing diverse points of view.

9 "*Un poëte contemporain a caractérisé ce sentiment de la personnalité de l'art et de l'impersonnalité de la science par ces mots: l'art, c'est moi; la science, c'est nous:*" Claude Bernard (quoting Victor Hugo), *Introduction à l'étude de la médecine expérimentale* (Paris: 1865), ch. II, § IV.

10 "Recording" covers a lot, including long-standard *sound* recording on cinema film by optical means. I discuss this in Maynard (2007: 327–8). Some may wish to distinguish from "mere" reproduction another photo-function: instantaneous distribution to masses of reception sites from central sources: *broadcasting*.

11 Lest this seem to be only incunabula, as noted below, computer chips are printed by photographic methods remarkable akin to Niépce's photo-etching (neglected by Arago), right down to the material, whereas daguerreotypy proved a dead end (see Maynard 2005: 118).

12 Walter Benjamin's 1936 essay (properly translated "The Work of Art in the Age of Mechanical Reproducibility") is available online and in various printed editions in several languages. This and the following observations on Herschel's cyanotype are due to Ward (1998).

13 For a developed theory of depictive representation in terms of imagining, including self-imagining see Walton (1990: ch. 8).

14 See P. Felsch, "Marey's Flip Book" online essay http://vlp.mpiwg-berlin.mpg.de/essays/data/art31?p=1.

15 Hubble's famous discoveries are of course based on the "real candle" discovery of Henrietta S. Leavitt, head of photometry at Harvard, whose laborious comparisons of photographic plates had established the luminance/variable function for Cepheid variables.

16 For more details, see http://solarsystem.nasa.gov/galileo/index.cfm.

17 Note on sonification: "One approach scientists use to make sense of the data from instruments is to make pictures and graphs to represent the data ... 'data visualization.' Some types of data, especially radio signals, are very similar in many ways to sound. The power of a radio signal is analogous to the volume of a sound. The radio signal also varies in terms of the frequency and wavelength of the radio waves ... like the variation in pitch of sound waves. So scientists sometimes translate radio signals into sound to better understand the signals. This approach is called 'data sonification' ... On June 27, 1996, the Galileo spacecraft made the first flyby of Jupiter's largest moon ... The Plasma Wave Experiment (PWS), using an electric dipole antenna, recorded the signature of a magnetosphere at Ganymede ... the first example of a magnetosphere associated with a moon. The PWS data are represented here as both sounds and a rainbow-colored spectrogram. Approximately 45 minutes of PWS observations are transformed and compressed to 60 seconds. Time increases to the right and frequency (pitch) increases vertically. Color is used to indicate wave intensity ... The audio track represents the PWS data and is synchronized with the display of the rainbow-colored spectrogram. The pitch of the sound is reduced by a factor of nine from the measured frequency and follows the location of the signal on the rainbow-colored spectrogram." NASA, "Sounds of Jupiter," at https://solarsystem.nasa.gov/galileo/sounds.cfm.

18 See at https://solarsystem.nasa.gov/galileo/sounds.cfm; http://hubblesite.org/gallery/.

19 This was regarding an Olympic qualifying race in Eugene, Oregon. Marathon courses provide another interesting challenge for measurement, given that runners cannot traverse quite the same distances over twisting courses on broad, sometimes crowded tracks.

20 One of the clearest of several online accounts of this system is "A Camera Fast Enough to Watch Light Move?" *Skull in the Stars* website: http://skullsinthestars. com/2012/01/04/a-camera-fast-enough-to-watch-light-move/.

21 Stillman Drake's translation from a letter by Kepler (see Drake and Swerdlow 1999: 347), which I have slightly rearranged for rhyme.

References

Arago, D. F. 1839. Report to the French Academy of Sciences. In: *Classic Essays on Photography*, edited by A. Trachtenberg [1980], 15–25, New Haven, CT: Leete's Island Books.

Benjamin, W. 1936. *The Work of Art in the Age of Mechanical Reproducibility*, German edition online: https://archive.org/details/DasKunstwerkImZeitalterSeiner TechnischenReproduzierbarkeit, accessed March 14, 2016.

Bernard, C. 1865. *Introduction à l'étude de la médecine expérimentale*. Paris: Éditions Garnier-Flammarion.

Braun, M. 1994. *Picturing Time: The Work of Etienne-Jules Marey, 1830–1904*. Chicago: University of Chicago Press.

Daston, L. and Galison, P. 2007. *Objectivity*. Cambridge, MA.: MIT Books.

Daston, L. and Galison, P. 1992. The Image of Objectivity. *Representations 40 (Fall)*: 81–128.

Drake, S. and Swerdlow, L. (eds.) 1999. *Essays on Galileo and the History and Philosophy of Science I*. Toronto: University of Toronto.

Hubble, E. P. 1920. *Photographic Investigations of Faint Nebulae*, Publications of the Yerkes Observatory, vol. IV, pt. II. Chicago: University of Chicago Press.

Hunt, R. 1854. *A Manual of Photography*. 4th edn, London and Glasgow: Richard Griffin.

Ivins, W. M. Jr. 1953. *Prints and Visual Communication*. Cambridge, MA: MIT Press.

Layden, T. 2012. Photo finish examiner details how trials race was deemed a dead heat. *Sports Illustrated* (July 27–August 12).

Kandel, E. 2012. *The Age of Insight: The Quest to Understand the Unconscious in Art, Mind, and Brain, from Vienna 1900 to the Present*. New York: Random House.

Marey, E. J. 1878. *Méthode graphique dans les sciences experimentales*. Paris: G. Masson.

Maynard, P. 2007. We Can't, eh, Professors? Photo Aporia. In: *Photography Theory*, edited by J. Elkins, 319–33, New York: Routledge.

Maynard, P. 2005. *The Engine of Visualization*. Ithaca, NY: Cornell University Press.

Poe, E. A. 1840. The Daguerreotype. *Alexander's Weekly Messenger* (15 January) In: *Classic Essays on Photography*, edited by A. Trachtenberg [1980], 37–8, New Haven, CT: Leete's Island Books.

Plato, *Theaetetus*, http://www.gutenberg.org/files/1726/1726-h/1726-h.htm, accessed March 14, 2016.

Trachtenberg, A. (ed.). 1980. *Classic Essays on Photography*. New Haven, CT: Leete's Island Books.

Walton, K. 1990. *Mimesis as Make-Believe: On the Foundations of the Representational Arts*. Cambridge, MA: Harvard University Press.

Ward, J. 1998. John Herschel's Cyanotype: Invention or Discovery? *History of Photography* 22 (Winter): 371–9.

Internet sources:

Felsch, P. Marey's Flip Book. online essay http://vlp.mpiwg-berlin.mpg.de/essays/data/art31?p=1, accessed March 14, 2016.

NASA, Sounds of Jupiter. https://solarsystem.nasa.gov/galileo/sounds.cfm, accessed March 14, 2016.

NASA. Galileo. http://solarsystem.nasa.gov/galileo/index.cfm, accessed March 14, 2016.

NASA. Hubble. http://hubblesite.org/gallery/, accessed March 14, 2016.

Skulls in the stars. 2012. A Camera Fast Enough to Watch Light Move? http://skullsinthestars.com/2012/01/04/a-camera-fast-enough-to-watch-light-move/, accessed March 14, 2016.

5 The Media Aesthetics of Brain Imaging in Popular Science

Liv Hausken

Brain imaging technologies have been enthusiastically embraced not only by the scientific and medical community but also by the public, whose awareness of them is fed principally by popular media reports of neuroscientific methods and findings (see Weisberg 2008: 51–6). The images they help to generate are regarded as being epistemologically compelling (see Roskies 2007: 860–72) and are believed to have a particularly persuasive influence on the public perception of research on cognition (see Dumit 2004; McCabe and Castel 2008: 347–52; Weisberg 2008; Keehner et al. 2011: 422–8). However, a number of empirical studies have questioned the widespread assumption of the persuasiveness of brain images (see Michael et al. 2013: 720–5).

My aim in this contribution is critically to examine this notion of persuasiveness and the cultural assumptions that underpin it. To be able to discuss the complexity involved in the production of authoritative knowledge (see Joyce 2008) in relation to brain images I believe we need to consider scientific visualization in its social, cultural and historical context and with a particular emphasis on the visual cultures of which it is part. More specifically, I seek to introduce a media aesthetic perspective on brain images used in popular science accounts of neurological research (see Hausken 2013: 29–50). Media aesthetics is in the basic sense a humanistic endeavor. In media aesthetics, aesthetics is viewed not as a philosophy of art but rather as a theory of culturally and historically embedded sensation and perception.[1] The human perceiver is considered to be embedded in the sociocultural environment and interacts with it continuously in an engaged and multisensory fashion. This general model of aesthetic engagement is just as applicable to works of art and popular culture as it is to the built and natural environment. Hence, aesthetics here is not confined to a particular kind of object (such as art). Neither is it characterized by the specific properties of the object of inquiry. Rather, it is defined by the perspective through which the objects are approached. Perspective includes perceptual engagement as well as the influence of conceptual information and the ways in which conceptual knowledge may direct our perceptual scale and framing of the object in question.

Aesthetics is regarded here as a means of reflecting critically on cultural expressions, on technologies of the senses and on the experiences of everyday life. Further, *media* are considered not primarily in terms of the mass media or other social institutions and cultural formations but rather as very specific technological arrangements that can be identified as such by the way they activate experiences using a variety of techniques. In this conception of media, particular objects, situations or phenomena are studied as complex expressions of mediation and are regarded as tools for investigating cultural (pre)conditions and theoretical assumptions. The plurality of media is of interest not so much as a collection of narrowly technical entities or systems but rather as a reservoir of different technical premises, semiotic systems, modes, genres, and stylistic conventions, as well as of scholarly interests, academic discourses, and kinds of knowledge.

In a nutshell, then, the basic research interests represented by the field of media aesthetics concern the relationships between culturally and historically situated perception, media technologies and mediation (understood both as dissemination and, more fundamentally, as the idea that all meaning is conveyed through culture). In proposing a media aesthetic approach to brain imaging I am also keen to stress the importance of opening up the domains of science and art to encompass their general socio-cultural history, with particular attention to visual culture. To understand people's experiences and perceptions of brain images in the popular press it is also crucial to take into account perspectives from the humanities that are so strikingly lacking in studies of brain imaging.

Previous Studies of Images in Popular Neuroscience

First of all, the status of brain imaging must be seen in the context of the general status enjoyed by neuroscience in contemporary industrialized societies. A few years ago, Deena Skolnick Weisberg and her colleagues presented a study which concluded that neuroscientific explanations as such seem to hold a seductive fascination for nonexperts. Perhaps not surprisingly, subjects in three different test groups judged good explanations of psychological phenomena to be more satisfying than bad ones. However, subjects in nonexpert groups additionally judged explanations containing logically irrelevant neuroscience information to be more satisfying than explanations without such information. The neuroscience information had a particularly striking effect on the judgment of nonexperts concerning bad explanations, masking the otherwise salient problems in these explanations (see Weisberg et al. 2008: 470–7).

It seems plausible to expect that technical language comes across as more convincing for lay people. However, this study suggests a closer affinity between the findings and what has been called the "seductive details effect" (Weisberg et al. 2008: 476); that is, related but logically irrelevant details presented as part of the argument attract the reader's attention and divert it from the general argument of the text. Another key to the seductive allure of neuroscience explanations of psychological phenomena suggested in this study is the assumption that these explanations appeal to people's tendency to accept reductionist explanations: "Because the neuroscience explanations in the current study shared this general format of reducing psychological phenomena to their lower-level neuroscientific counterparts, participants may have jumped to the conclusion that the neuroscience information provided them with a physical explanation for a behavioral phenomenon. The mere mention of a lower level of analysis may have made the bad behavioral explanations seem connected to a larger explanatory system, and hence more insightful" (ibid.).

Although their study does not include the possible impact of brain images on people's perceptions of research on cognition, Weisberg and her colleagues do suggest that the association of neuroscience with powerful visual imagery may be significant in this context. Visual imagery may merely attract attention to neuroscience studies. Referring to a study by David P. McCabe and Alan D. Castel that was in press at that time, they argue that visual imagery is known to render scientific claims more convincing. My question is: if this is so, why?

In *Seeing is Believing: The Effect of Brain Images on Judgments of Scientific Reasoning* (2008), David P. McCabe and Alan D. Castel report on three experiments which show that presenting participants with brain images that accompany articles summarizing cognitive neuroscience research resulted in higher ratings of the scientific reasoning presented to support the arguments made in those articles than presenting them with articles accompanied by bar graphs, a topographical map of brain activation or no image at all.[2] The primary purpose of the study was to examine whether brain images actually *do* have a particularly powerful persuasive influence on the perceived credibility of cognitive neuroscience data (see McCabe and Castel 2008: 344). Their findings confirm that they do (see ibid.: 350). Bar graphs were chosen because they are generally regarded as a particularly effective way to communicate scientific data and because brain imaging data are, in fact, often presented using bar graphs in research articles. However, since bar graphs are quite simple, a topographical map of brain activation was included in the study to address the possible impact of visual complexity on judgments of scientific reasoning (see ibid.: 347).[3]

If visual complexity is of major importance in this context, there should be no major difference between the topographical map and the brain image in the experiment. The experiment nevertheless showed that the reporting of research including brain images resulted in higher ratings of the scientific reasoning used to express the arguments made in the articles compared to the reasoning used in articles containing other sorts of illustrations. McCabe and Castel state rather casually that topographical maps "are not typically used in the popular press and presumably are not as easily identified as representing the brain" (ibid.: 347). Equally casually, they suggest that: "[b]rain images may be more persuasive than other representations of brain activity because they provide a tangible physical explanation for cognitive processes that is easily interpreted as such" (ibid.: 349). I find their tentative conclusions suggestive but insufficiently explained.

In their study, McCabe and Castel do not seem to have drawn the participants' attention directly to the images as such. They only ask how convincing the participants find the arguments put forward in the articles, without inviting them to comment on the images (or the absence of images) contained therein. This is precisely what Madeleine Keehner, Lisa Mayberry and Martin H. Fisher do in their article entitled *Different Clues from Different Views: The Role of Image Format in Public Perceptions of Neuroimaging Results* (2011). In the first of two experiments they focused on perceptions of nonexperts on five different types of brain images, referred to as a "whole brain image," an "inflated brain," an axial "brain slice," a "glass brain' and a "topographic map" (Keehner et al. 2011: 423). The participants were informed that these represented different types of brain images. They were then asked to rate the five images according to perceived complexity, resemblance to a real brain and apparent three-dimensionality (see ibid.).[4]

In the second experiment, first year undergraduate psychology students were asked to rate the scientific reasoning and the quality of explanations in five articles, each of them paired with one of the five brain images from the first experiment. Thereafter, the five images were displayed without the texts for the participants to rate for image complexity and image familiarity (see ibid.: 424).

The experiments in Keehner et al. were based on two hypotheses, the first being that brain images are convincing because of their perceived complexity (see ibid.: 422). In doing so, they take the findings reported by McCabe and Castel as a starting point but dispute their conclusion. Where McCabe and Castel found no supporting evidence for the complexity hypothesis, Keehner et al. show that perceived image complexity has only a weak effect on text credibility. Admittedly, and as the authors also state, measuring perceived complexity poses a considerable challenge (see ibid.: 427), so nothing conclusive can be said with regard to image complexity in this context.

The other hypothesis on which the experiments were based is that "brain images are more persuasive because of their perceived *realism*" (ibid.: 423). With reference to Harvey S. Smallman and Mark St. John's article *Naive Realism: Misplaced Faith in Realistic Displays* (2005), Keehner et al. distinguish between visual resemblance to a real physical object (in this case, a brain) and realism of represented space, explained as displays "where pictorial depth cues such as shading and perspective-view give the impression of *three-dimensionality*" (Keehner et al. 2011: 423). For the second experiment, participants were asked separately to report their familiarity with each image type in order to establish "whether more familiar brain images would produce more favorable ratings of information associated with them" (ibid.: 423).

For familiarity ratings, two of the five image types stood out clearly as much more familiar to the nonexperts than the others, namely the image of the whole brain and of the brain slice (in contrast to the participants' low familiarity with topographical brain maps, images of inflated brains and of glass brains (see ibid.: 426)). However, the experiment established no difference in credibility ratings for articles including (or not including) the most *familiar* brain images. Still, the researchers' findings show that the whole brain image as well as the image of the inflated brain made accompanying texts more convincing than texts paired with other images. According to this study, the main factor distinguishing these two images from the other ones was their *perceived three-dimensionality* (see ibid.: 426). Both the images of the whole brain and of the inflated brain (which is similar in many respects, but with the sulci, or furrows, pushed outward) contain depth cues that convey an impression of three-dimensionality (see ibid.: 423). This is an interesting result and one that the authors also attempt to reflect upon (see ibid.: 426–7).

Why should the apparent three-dimensionality of a brain image cause nonexperts to allocate higher credibility ratings to associated texts? Keehner et al. suggest that it relates to "how we conceptualize different kinds of displays" (ibid.: 426). The perceived three-dimensionality may have contributed to a general impression of concreteness or solidity, which may possibly have encouraged the participants to consider the 3D images as "direct *depictions*, representing something tangible or visible, rather than nonfigurative graphics. … Such apparent (but not actual) directness may have increased *processing fluency* or *intellectual fluency*, which is the feeling that one understands and can extract information with relative ease" (ibid.: 426). I find these reflections better developed and more nuanced than those presented in the concluding remarks made by McCabe and Castel. Nevertheless, what I seek to show is that debates about the role of brain images in public perceptions of research on cognition would benefit from more knowledge about scientific visualization as part of visual culture and socio-cultural history.

The major contribution to this debate of Keehner et al. is their focus on brain images as such and their insistence that the choice of image format matters (see ibid.: 422–6). Where McCabe and Castel talk about brain images as if they constitute a distinct category of images, the five types of brain images specified in Keehner et al.'s study help to broaden the scope of the discussion and generate further insight. Despite this difference, McCabe and Castel's conclusions are strikingly similar to Keehner et al.'s study: "Brain images may be more persuasive than other representations of brain activity because they provide a tangible physical explanation for cognitive processes that is easily interpreted as such" (McCabe and Castel 2008: 349).

The illustrated examples of brain images given by McCabe and Castel are those referred to by Keehner et al. as the most familiar types, that is, "a whole brain" and "a cross sectional brain slice" (see ibid.: Fig. 1a and Fig. 2a.). But they do not attempt to explain why these particular brain images should support the idea that the tangible and physical can explain cognitive processes in popular accounts of neuroscience. In emphasizing the impression of concreteness as being capable of depiction and of depictions as being easily understood, Keehner et al. introduce into the debate an important element of reflection on conceptions of visual norms. Still, it is by no means obvious why an impression of solidity and concreteness should be considered easier to depict and thus (so the argument goes) easier to understand than visual similarity and recognizability arising from the familiarity of an image.

The impression of concreteness pointed out by Keehner et al. refers to one of two properties of aesthetic realism—realistic *space*—as opposed to "realism of represented *objects*—visual similarity or resemblance between images in the display and the real objects they represent" (Keehner et al. 2011: 423). However, the idea that the impression of concreteness is a function of a realistic representation of space, rather (or more) than of a realistic representation of the object, is most peculiar. Further, it is not obvious why a realistic representation of an object should be synonymous with *visual* resemblance. Without entering into a detailed and sophisticated discussion of aesthetic realism, suffice it to say that the issue is a question of belief, a general attempt to present something as believable; something is presented (socially, emotionally, visually or otherwise) so that it appears to be real. It is hard to see how visual realism should be different in principle from spatial realism in this respect.

Finally, before bringing this discussion of the importance of realism in the above studies to a close, I would like to question the assumption that nonexperts can assess the visual resemblance between brain images and real brains given that their access to real brains is limited to the images displayed in popular neuroscience texts. Moreover, although it may be valuable to include separate reports on image familiarity for the different image types in a study like this, it only takes into account the participants' active *recognition* of images, not images or other modes of expression they may be accustomed to without recalling them as such. We do not just *look at* or *look through* images; we see *according to* them, or *with* them, to paraphrase Maurice Merleau-Ponty (1964: 126). This seeing-with that is underscored by Merleau-Ponty has been underestimated in contemporary theories of visualization, which tend to focus excessively on images as representations. If images are not just to be looked at and then recognized but also to be seen with or according to, the relationship between realism and image familiarity referred to above turns out to be far more complex.

Visual Cultures of Brain Imaging in Popular Neuroscience

Regardless of the reasons for the importance of perceived concreteness in the above-mentioned studies, the image format seems to matter when it comes to public perceptions of research on cognition. In line with the media aesthetic perspective advocated here, the challenge is to pay attention to what seems to be taken for granted in a certain perception, that is, the sense of what is given (see Hausken 2013: 162). What is the taken-for-grantedness in the formats that are used in the public press? What are the cultural assumptions at play?

Not all types of scientific visualization of brain imaging data seem to count as brain images in the public perception of neuroscience. As we know from the studies mentioned above, in the scientific community brain imaging data are presented in a variety of ways, the most common probably being bar graphs and topographical maps (see McCabe and Castel 2008: 347). Beauty contests of popular medical science, such as the *Wellcome Image Awards*, seem to celebrate small-scale endeavors, such as images of cells of the nervous system.

In popular science and in the press, however, a rather limited range of visual forms and visualization techniques are chosen to illustrate reports of scientific discoveries related to the brain. A research article on Alzheimer's Disease (AD), for instance, may present the brain imaging data by using a bar chart comparing tissue from a selection of AD patients with control tissue. Since bar graphs are generally seen to represent a particularly effective way to communicate scientific data, one might expect popular science magazines to rely on the scientific authority of the bar graph to communicate research on brain related diseases. However, more often than not such cases are illustrated using an image of a slice of a brain from three different views, the profile (sagittal view), the frontal (coronal) view or the view from above (transaxial view), the latter being referred to as "a brain slice" in the studies above.

The brain imaging techniques or modalities indirectly referred to seem first and foremost to be different MR imaging techniques as well as CT and PET scans. Just as the visualizations of staining cells and the *Brainbow mouse* celebrated by scientific imaging awards involve very specific brain imaging

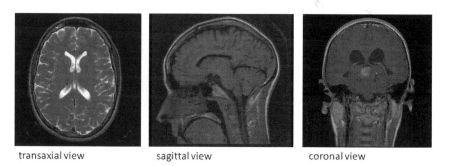

transaxial view   sagittal view   coronal view

Figure 5.1 The view of the brain from above (transaxial), the profile (sagittal view) and the frontal (coronal view, with strongly marked neck bone).

Sources: transaxial and sagittal view: own images, coronal view: Wikimedia Commons, http://commons.wikimedia.org/wiki/File:T1-weighted_coronal_MRI_image_post_contrast_tectal_plate_glioma.jpg, licensed under the Creative Commons Attribution-Share Alike 3.0 Unported http://creativecommons.org/licenses/by-sa/3.0/ license.

data, so too do the PET scan, the CT and the MRI. The PET scan, for instance, measures emissions from radioactively labeled chemicals that have been injected into the bloodstream and uses the data to produce images of the distribution of the chemicals throughout the brain. In other words, these brain imaging data are as technically specialized and confined in their depictions as the brain imaging data involved in the less well-known visualization techniques often awarded prizes in scientific imaging contests.

The question is, then, what are the cultural assumptions at play when visualizations involving CT, MRI or PET scans are so often presented in the public press as images of human brains? One could argue that since these views are typically used in the popular press, they will more easily be identified as representing the human brain (see McCabe and Castel 2008: 347). However, the question still remains: why are they typically used in the popular press? What is the taken-for-grantedness in these iconic images?

In *Picturing Personhood. Brain Scans and Biomedical Identity* (2004) Joseph Dumit suggests the following: "In its popular usage, a brain image is akin to the simplified reality of a graphic cartoon" (ibid.: 145). He refers to Scott McCloud's idea of comics as "a form of amplification through simplification that focuses our attention on an idea" (ibid.). In *Understanding Comics: The Invisible Art* (1993) McCloud uses the form and format of the graphic cartoon to explore the phenomenon of comics. On his way through this book-length reflection, McCloud's storyteller pauses "… to examine cartooning as a form of amplification through simplification" (McCloud 1993: 30). "When we abstract an image through cartooning, we're not so much eliminating detail as we are focusing on specific details," he explains in the next frame (ibid.). "The ability of cartoons to focus our attention on an idea is, I think, an important part of their special power …" (ibid.: 31).

Then Dumit leaves McCloud and his cartoon character. However, it is precisely at this point where McCloud's reflections become especially interesting and relevant for the argument being made here: "… I believe there's something more at work in our minds when we view a cartoon—especially of a human face—which warrants further investigation," he says (ibid.: 32). "The fact that your mind is capable of taking a circle, two dots and a line and turning them into a face is nothing short of incredible!" Thrilled by this, McCloud's storyteller proclaims: "But still more incredible is the fact that you cannot avoid seeing a face here. Your mind won't let you!" A few frames later, this ink figure stands next to a huge image of a power socket, as if he is giving a lecture about this picture, putting forth the argument: "We humans are a self-centered race" (ibid.).

McCloud does not refer to images of the brain. However, inspired by the novel comparison with cartoons suggested by Dumit, I want to propose some very specific features of the popular use of brain imaging that play a major part in public perceptions of research on cognition.[5]

First of all, visualizations of brain imaging data involving CT, MRI or PET scans seem easily translatable into views recognized as human due to their typical reference to volume, scale and shape. The impression of three-dimensionality stressed by the studies referred to above does not just contribute to a general impression of concreteness, solidity and objecthood (see Keehner et al. 2011: 426). Rather, more than this, the scanned object looks *complete*, autonomous. In a study of how human egg cells are visually displayed in public information about assisted reproduction, the anthropologist Merete Lie shows how the cells are presented as independent of human bodies. She argues that this disassociation from human bodies should be considered as part of a process of objectifying the cells as entities to be used and manipulated (see Lie 2012: 475–96).

Despite many differences between singular cells (like human egg cells) and complex organs (like brains), the impression of the brain scan as referring to a physical object can be said to be due not only to the perceived three-dimensionality but also to the appearance of an object as cut off from the environment: the brain appears to be alone, autonomous, self-sufficient. This impression of self-enclosure is reinforced by the common practice of combining several brain imaging technologies in a *compound* visualization.

The history of medical imaging is normally portrayed as an endless progression of seeing more and better (Keyles 1997). Various technologies are often combined to provide information concerning both structure (e.g. MRI) and activity (e.g. PET) (see Beaulieu 2000: 6), resulting in a very complex temporal object (see Dumit 2004: 195–6, n. 29–30) that is in conflict with the notion of a solid, physical object. Despite this, the practices of combining multiple tools to produce a complex image seem to underpin the popular perception of medical images as enabling us to see more in *the same* physical object (here, one particular brain).

But scale also matters. In contrast to both human egg cells (being about 100 micrometers in diameter) on the one hand and objects in the observable universe on the other (like the Earth, which is about 12,756.32 kilometres in diameter at the equator), the scale of the object in question is very much what one might call a *human-size scale*.[6] Most of us have seen plenty of images of human egg cells as well as images of the earth. However, very few have seen Planet Earth with their own eyes. Even the human egg cell, which is barely visible to the unaided eye even under the right conditions, is not an object people normally look at without using magnification. One may very well argue that we are similarly not familiar with living brains. I will nevertheless suggest that the size of the physical object referred to by the specific visualizations of brain imaging data discussed here (the three views mentioned above) plays a role in how images are perceived with reference to an object considered to be truly human.

Further, I will suggest that this apparent three-dimensional imagery is easily perceived not just as a complete object from the viewer's own world but as an image of a human shape. The semblance of the human profile (sagittal view), the quite uncanny appearance of protruding eyes in an image of a face without skin (coronal view), or a picture that cuts across left and right hemispheres (transaxial view) indicating a physical brain as removed during autopsy, all trade on people's familiarity with visual representations of the human head.

In other words, it is not just a "perceived three dimensionality" that may have contributed to the general impression of "concreteness or solidity" which, in turn, may have encouraged the participants in various experiments to consider 3D images as "direct *depictions*, representing something tangible or visible, rather than nonfigurative graphics," as suggested by Keehner et al. (see Keehner et al. 2008: 426). In addition to the impression of completeness, the references to *human size and human shapes* indicate the importance of understanding how a *human perspective* plays a role in popular perceptions of brain images.

In his delightful essay "Emergence," the scientist and philosopher Michael Polanyi (1966) compares the reality of things—such as cobblestones—with the reality of human matters:

> Persons and problems are felt to be more profound [than cobblestones], because we expect them yet to reveal themselves in unexpected ways in the future, while cobblestones evoke no such expectation. This capacity of a thing to reveal itself in unexpected ways in the future I attribute to the fact that the thing observed is an aspect of a reality, possessing a significance that is not exhausted by our conception of any single aspect of it. To trust that a thing we know is real is, in this sense, to feel that it has the independence and power for manifesting itself in yet unthought of ways in the future. I shall say, accordingly, that minds and problems possess a deeper reality than cobblestones, although cobblestones are admittedly more real in the sense of being *tangible*. And since I regard the significance of a thing as more important than its tangibility, I shall say that minds and problems are more real than cobblestones.
>
> (Polanyi 1983: 32–3)

Following Polanyi on this point, I suggest that if we consider the significance of human matters to be more important or real than the tangibility of physical objects, it is quite strange that leading research on popular perceptions of brain imaging emphasizes objecthood and tangibility to such a great extent, with little or no reflection about how these images appear to be images of something that is humanly real.

The Historicity of Cultural Forms

So far I have introduced perspectives from media aesthetics and, more broadly, from the humanities in order to supplement, differentiate and critique research on popular understandings of brain images. These perspectives are represented most explicitly by Maurice Merleau-Ponty with his conception of seeing according to images (or with them), Scott McCloud with his humanistic perspective on the history of cartoons and, finally, Michael Polanyi, who emphasizes the significance of things rather than their physical qualities. In the final part of this contribution I will suggest that in order to comprehend popular conceptions of brain imaging, one must also take into account one of the most basic discourses in the humanities which is often seemingly neglected in studies of contemporary science and technology, namely the discourse of historicity. To understand popular conceptions of contemporary brain imaging an awareness of the historical dimensions of both the object of study and the conception of theory is required.

Merete Lie makes us aware of a certain *practice of illustration* in governmental information about assisted reproduction (see Lie 2012). In a similar manner Joseph Dumit stresses the importance of looking at specific illustration practices to understand the popular impression of brain images as *easily understandable* (see Dumit 2004). Meanwhile Keehner et al. emphasize the impression of concreteness as encouraging the participants in their experiments to consider the 3D images as "direct *depictions*" and argue that such "apparent (but not actual) directness may have increased ... the feeling that one understands and can extract information with relative ease" (Keehner et al. 2011: 426). Their argument is based on the general idea that people wrongly take brain images to be photographs.

Without entering into a discussion of their rather problematic conception of photography, I want to underline how both brain images and photographs are always situated in specific contexts and are subject to various conventions. According to Dumit, scientific journals often use rather extreme or exemplary images of the brain to illustrate an argument. Their readers are familiar with this practice. However, when the same images are transferred to popular science magazines, their context and the captions accompanying them are altered so that they often appear to be representative images. As such, they convey the impression that we can see and understand brain images with ease (see Dumit 2004: 95–100). Regardless of the audience's perception of these images as photographs, this illustration practice thus helps to explain the idea that brain images are easily understood.

In the following I will suggest that a particular conception of the photographic portrait plays an important part in public perceptions of brain imaging. It is commonly argued (by way of critique) that "[n]on-specialists think of neuroimages as analogues to photographs of the brain" (Roskies 2007: 861)[7] whereas, of course, they are not. My argument here, however, is of a different nature. I have argued that a particular range of techniques and a limited selection of views characterize the use of brain images in the popular press. Furthermore, I have suggested that human shape and scale along with an impression of volume serve to reinforce the idea of a complete image of the human head into which we can look with the help of ever better technologies that are often combined to reveal even more that we can understand with ease, thanks to the practices of compound imaging and exemplary illustrations. Just as McCloud's cartoon character proclaims that we are not only capable of taking a circle, two dots and a line and turning them into a face but that indeed we cannot avoid doing so, I want to suggest that the use of brain images in the popular press relies on a few quite specific photographic genres which are so established and common in Western culture that we hardly notice them as conventional.

I have already suggested above that the three common brain imaging views widely used in the popular media (the sagittal view, the coronal view and the transaxial view) all trade on people's familiarity with visual presentations of the human head. More specifically, the latter may seem to show a brain cut through and is therefore most likely to be associated with pathology, while the first two appear to depict a living individual by virtue of their formal similarity to different genres of the portrait, such as the ID portrait and the "mugshot." Thus, if we follow Merleau-Ponty and look at brain images not as representations of objects or persons but rather as images to be seen with or according to, I will suggest that the sagittal (profile) and coronal (frontal) views in particular are not seen simply as representations of a brain but rather *according to* portraits of human beings.

One may object that, in contrast to portraits, brain images are not images of individual brains. As Weisberg notes:

> pictures of brain activation in the popular press are not pictures of single brains, as people unfamiliar with fMRI technology might assume. Instead, in sharp contrast to photographs, these pictures provide summaries of the activation patterns from several brains.
>
> (Weisberg 2008: 52)

Although this may be correct for brain images "in the popular press," where the journalist will typically be interested in illustrating general arguments about specific diseases or conditions, the neuroimaging techniques used in clinical practice for diagnostic reasons indicate that brain imaging per se cannot be defined as imaging across individuals. A parallel argument can be made for photographic portraits: in the 19th century, composite photography emerged as part of the development of social statistics (see Sekula 1986). When several separate photographs were superimposed on top of one another, the resulting composite image indicated an average, or a type. In other words, photographic techniques and brain imaging techniques alike may be used to display both individuals and types.

However, if we concern ourselves only with the most common and well-known brain images and photographic portraits, Weisberg is, of course, right. Publicly displayed brain images are often averaged and, as Roskies points out, "the resulting image is not only a generalization over time, but also a generalization over individuals" (Roskies 2007: 870). Equally, however, according to the standard definition a portrait serves as a presentation of a person considered for him- or herself—or, as Jean-Luc Nancy puts it rather more specifically, "to produce the exposition of the subject" (Nancy

2006: 222). In order to do so, however, a whole range of social and cultural norms need to be taken into consideration. The relationships and tensions that exist between systems of collective and individual registration are closely interlinked with one another, as are those between identity and identification. As Jane Caplan and John Torpey stress in the introduction to their extensive collection, *Documenting Individual Identity* (2001), the question "who is this person?" leaches constantly into the question "what kind of person is this?". "Identification as an individual is scarcely thinkable without categories of collective identity ... " (Caplan and Torpey 2001: 3).

For portraits, nowhere is this more glaringly apparent than in identification photos—bureaucratic images that transform portraits into efficient instruments for identification and discipline by archiving and storing them in databases. Police archives and medical atlases make it possible to compare data from a multitude of legal and medical sources in order, on the one hand, to identify criminals and, on the other, to elucidate average or typical phenomena across multiple instances. In other words, seeing brain images with (or according to) ID photos or mugshots is not as strange as it may seem. Mugshots belong to the genre of ID photos and contribute to the widespread perception that ID photos depict a person as a criminal.

Compared with the formal signs of bourgeois self-presentation in the 19th century (such as the three-quarter profile and the emphasis on the face and hands), both portraits demonstrate how the formal composition of the image is linked to questions of race, social status and power. Yet the bourgeois portrait itself also begs the question of what kind of person this is. As many have argued, such features of the portrait owe their continuing signifying power to the physiognomic discourse with which nineteenths century portrait photography was deeply enmeshed and which continues to linger more than one century later (see e.g. Long 2007: 53–5). In this sense, the use of brain images in the popular press should be seen not just in relation to the 19th century photographic portrait but also to contemporary ideas of physiognomy.

In the 19th century, phrenology and physiognomy were two interrelated pseudosciences motivated by the desire to classify bodies according to their visual appearance (see e.g. Sekula 1986). Phrenology focused primarily on measurements of the human skull, based on the concept that the brain is the organ of the mind and that certain brain areas have localized specific functions. Physiognomy, on the other hand, was the art of judging human character from physical appearance, particularly the face. Of the two, physiognomy seems to have been far more influential upon visual arts practice than phrenology (see Freeland 2010: 125). Both of these ways of seeing have been accused of being politically problematic (e.g. racist (see Sekula 1986)).

Brain imaging of today has nevertheless acknowledged its proximity to phrenology (see e.g. Dumit 2004: 23–4). Even if the use of brain images in popular science seems to rely on the idea that the physical brain can easily express a person's mental state just as the face was supposed to reveal her character, there seems to be no acceptance of historical heritage when it comes to physiognomy. This reference to the politics of 19th century physiognomy is not meant as an indictment of the contemporary case of brain imaging. Rather, the aim of underlining the historicity of scientific imaging is to highlight the need for a careful analysis of the visual practices of brain imaging and the histories of mediated forms and their social and cultural discourses.

* * *

In cognitive brain imaging the aim is the better to understand psychological processes. In popular perceptions of research on cognition, brain images are often presented as pictures that provide physical explanations for cognitive phenomena. It seems reasonable to assume a correlation between physical and psychological conditions (see e.g. Weisberg 2008: 53). But what are the conditions for believing that pictures of the brain can provide physical explanations for cognitive phenomena? In this contribution, I have introduced a media aesthetics perspective and have highlighted some basic terms from the humanities to show that, in order to understand popular conceptions of brain imaging, one must not just consider what participants in an experimental setting report about convincing arguments or recognizable imagery but must also critically examine the experimental set up itself (including the image format chosen for the experiment), as well as the cultural assumptions at play.

I have argued that popular conceptions of brain imaging cannot be fully explained by the general tendency to "biologize" the mind in contemporary cognitive sciences. Instead, they need to be seen as part of specific visual cultures and viewed in the context of the historicity of cultural forms. Finally, in order to understand popular conceptions of brain imaging it is absolutely vital to adopt a humanistic perspective on the subject matter, that is to emphasize the agency of human beings and the role of meaning-making in how people perceive the images they see.

*A brief version of this contribution was first presented at the ESF-LIU Conference *Images and Visualisation: Imaging Technology, Truth and Trust*, 17–21 September 2012, Norrköping, Sweden, and then contextualized in a broader paper on photographic measurements at the Dimensions of Measurement Conference at the Center for Interdisciplinary Research, Universität Bielefeld, March 14–16, 2013. I am grateful for the professional interest and constructive feedback given at both these venues, as well as from the two research groups *Media Aesthetics* at the University of Oslo and *Images of Knowledge* at the University of Bergen.

Notes

1 In media aesthetics, aesthetics is developed conceptually from the original Greek sense of the term *aisthesis*, or sense perception.
2 In McCabe and Castel's study, concepts of "image" and "representation" appear to be used synonymously and include "topographical maps" (referred to inter-

changeably as "images" and "representations"). Thus a possible conception of verbal representation is excluded. More importantly, this blurring of the notions of *image* and *representation* makes it difficult to implement concepts of image other than those dominated by the paradigm of representation. I return to this later on.

3 It should be noted that brain imaging technologies do not measure brain activation directly. As underlined by Weisberg (2008) and others, "What fMRI scanners actually measure—and only indirectly at that—is the amount of blood flow to a given brain area, a reliable correlate of neural activity." Scientists know this, but "there is a danger that most consumers of fMRI pictures believe them to be something which they are not: a direct picture of activity in a single brain" (Weisberg 2008: 52). See also Roskies (2007: 864) and Dumit (2004: 71).

4 It is worth noting that here "the topographic map" (unlike in the study by McCabe and Castel) is treated as a brain image. However, in their illustration of brain images (Fig. 1) Keehner et al.'s topographic map appears as a cross between a graphic map and a sketch of the outline of a brain in profile and is therefore less schematic than the topographic map shown in McCabe and Castel (2008: 348, Fig. 2).

5 This particular part of the argument is presented in a brief essay in *Leonardo: Journal of the International Society for the Arts, Sciences and Technology*, 48:1 (January 2015), or online as AOP [Advance Online Publication], http://www.mit pressjournals.org.pallas2.tcl.sc.edu/toc/leon/0/ja.

6 "Human scale" refers here simply to the set of physical qualities and quantities of information characterizing the human body and its motor, sensory or mental capabilities. It also refers to human social institutions.

7 See also Keehner et al. 2011.

References

Beaulieu, A. J. 2000. New Views of the Mind-in-the-Brain. ch.1 from *The Space Inside the Skull. Digital Representations, Brain Mapping and Cognitive Neuroscience in the Decade of the Brain*. Amsterdam University. http://www.virtualknowledgestudio.nl/staff/anne-beaulieu/New-Views- of-the-Mind-in-the-Brain.pdf, accessed October 1, 2016.

Caplan, J. and Torpey, J. 2001. *Documenting Individual Identity. The Development of State Practices in the Modern World*. Princeton: Princeton University Press.

Dumit, J. 2004. *Picturing Personhood. Brain Scans and Biomedical Identity*. Princeton: Princeton University Press.

Freeland, C. 2010. *Portraits and Persons. A Philosophical Inquiry*. Oxford: Oxford University Press.

Hausken, L. 2013a. Introduction. In: *Thinking Media Aesthetics. Media Studies, Film Studies, and the Arts*, edited by L. Hausken, 29–50, Frankfurt/Main: Peter Lang Academic Publisher.

Hausken, L. 2013b. Doing Media Aesthetics: The Case of Alice Miceli's *88 from 14.000*. In: *Thinking Media Aesthetics. Media Studies, Film Studies, and the Arts*, edited by L. Hausken, 161–88, Frankfurt/Main: Peter Lang Academic Publisher.

Joyce, K. A. 2008. *Magnetic Appeal. MRI and the Myth of Transparency*. Ithaca, NY: Cornell University Press.

Keehner, M., Mayberry, L., and Fisher, M. H. 2011. Different Clues from Different Views: the Role of Image Format in Public Perceptions of Neuroimaging Results. *Psychonomic Bulletin & Review* 18:2: 422–8.

Kevles, B. 1997. *Naked to the Bone*. New Brunswick, N.J.: Rutgers University Press.

Lie, M. 2012. Reproductive Images: The Autonomous Cell. *Science as Culture* 21:*4*: 475–96.

Long, J. J. 2007. *W. G. Sebald: Image, Archive, Modernity*. New York: Columbia University Press.

McCabe, D. P. and Castel A. D. 2008. Seeing is Believing: The Effect of Brain Images on Judgments of Scientific Reasoning. *Cognition* 107:*1*: 347–52.

McCloud, S. 1993. *Understanding Comics: The Invisible Art*. New York: Harper Perennial.

Merleau-Ponty, M. 1964. *L'Œil et l'esprit*. Paris: Editions Gallimard.

Michael, R. B. et al. 2013. On the (non)persuasive Power of a Brain Image. *Psychonomic Bulletin & Review* 20:*4*: 720–5.

Nancy, J.-L. 2006. *Multiple Arts. The Muses II*. Redwood City, CA: Stanford University Press.

Polanyi, M. 1983. *The Tacit Dimension*. Gloucester, MS: Peter Smith.

Roskies, A. L. 2007. Are Neuroimages Like Photographs of the Brain? *Philosophy of Science* 74: 860–72.

Sekula, A. 1986. The Body and the Archive. *October* 39: 3–64.

Smallman, H. S. and St. John, M. 2005. Naive Realism: Misplaced Faith in Realistic Displays. *Ergonomics in Design: The Quarterly of Human Factors Applications* 13:*3*: 6–13.

Weisberg, D. S. 2008. Caveat Lector: The Presentation of Neuroscience Information in the Popular Media. *The Scientific Review of Mental Health Practice* 6:*1*: 51–6.

Weisberg, D. S., Keil, F. C., Goodstein, J., Rawson, E., and Gray, J. R. 2008. The Seductive Allure of Neuroscience Explanations. *Journal of Cognitive Neuroscience* 20:*3*: 470–7.

6 Compressed Sensing—A New Mode of Measurement

Thomas Vogt

Introduction

Making digital representations of physical objects has been approached with a pessimistic attitude demanding a very high rate of regularly-spaced measurements without taking into account that the object itself might have sparsity. In this text sparsity is used as an operational gauge of an object's complexity rather than a well-defined mathematical property. In mathematics we define a sparse matrix in contrast to a dense one as containing mostly zeroes. Compressed sampling takes into account the sparsity of an object and is able to successfully reconstruct images even after dramatically reducing the number of measurements required without loss of reconstruction fidelity. If one defines "sampling" as the act of performing measurements of different objects such as pixels in an image, one can conservatively measure one at a time or group several objects and measure groupings. Compressed sensing allows us to optimize measurements of such groupings and thereby perform significantly fewer measurements. This is related to the 12-ball problem, where one is tasked with finding one lighter or heavier ball out of a set of 12 balls by only three comparative weighing of groupings.[1]

Compressed sampling will not replace conventional sampling but complement it by making measurements possible which before were prohibitively costly. Cost in this terminology refers to measurements that are too expensive, take too long, require too much energy or subject the object to damage when measuring in the conventional linear fashion.

Shannon-Nyquist Measurements and Compressed Sensing

Harmonic analysis has shown that signals we measure can be described as convolutions of mathematical basis functions with frequency and intensity as variables. The best known example is the *Fourier* series, where a signal can be decomposed into a sum of simple oscillating functions such as sines, cosines or complex exponentials multiplied with a weighting function. Signals can then be represented by their Fourier coefficients. Mathematics provides us a plethora of basis functions such as wavelets, ridgelets, curvelets, and contourlets to approximate signals efficiently.[2] The conventional approach when measuring signals such as sounds and images relies on Shannon's theorem which states that the sampling rate with which one should record or image must be at least twice the maximum frequency present in the signal in order to record, transmit and reconstruct with high fidelity. This frequency is called the *Nyquist* rate (Nyquist 1928: 617–44) and the *Shannon-Nyquist* theorem (Shannon 1949: 10–21) assures us that our sampling is dense enough to allow us to reconstruct original analogue signals such as a Coltrane saxophone solo or a lake view.

This immensely powerful theorem is used in most consumer audio and video electronics, in conventional analogue-to-digital (ADC) converters, and in medical imaging, i.e. ultrasound and magnetic resonance imaging (MRI). The latter points to the importance of the quality of reconstruction since overlooking or altering even minute features in images might have important consequences as they are often the goal of such measurements. Intuitively one would think that measuring faster and omitting data from the reconstruction will always result in a less faithful reconstruction of the original signal and should therefore be avoided at all cost. However, as I will show below, the amount of data that needs to be measured, stored and transmitted impacts the feasibility of a measurement and is often constrained by external "cost functions" such as the speed with which one can image and reconstruct and therefore actively control a process, the time a patient is scanned in a computer-aided tomography (CAT) or MRI procedure or the amount of energy needed for a measurement, to name just a few. These cost functions drove the exploration of new modes of measurement to circumvent the shortcomings mentioned above and others discussed later. Compressed sensing allows measurements that optimize external cost functions by measuring faster and less without a commensurate loss of reconstruction fidelity.

In the *Shannon-Nyquist* measurement mode we uniformly and linearly measure our signal often by sequential line scans ("raster scans"). This results in many data points N, as we all realize when confronted with the storage capacities of our digital cameras. Therefore, we use programs based on compression algorithms which extract a subset K<<N that then stores our soundscape as MPEG and our images as JPEG files. Compression is a nonlinear process and relies on the fact that many of the Fourier or wavelet transform coefficients with which we can describe our signals are small or close to zero. In the case of JPEG or JPEG-2000 the signal is represented in a mathematical basis space different than the pixel basis, relying on the fact that most images have many zeroes in the discrete cosine (JPEG) or wavelet (JPEG-2000) basis.

The conceptual leap compression algorithms are based on is that we do not need to store zeroes or minute coefficients in order to be able to reconstruct our signal to a certain degree of fidelity. This shows that not all frequencies and all pixels or voxels have the same importance when we reconstruct the original signal. Some are more important than others and sparsity might provide us with an opportunity to escape the constraints of the *Shannon-Nyquist* theorem and reduce the number of coefficients we need to store and be able to reconstruct our original signal. In other words, some data sets that we measured uniformly at the *Nyquist* rate turn out to be sparse. We can therefore compress them for data transmission and storage purposes. Sparsity is thus a working measure of the complexity or lack thereof in a signal representation. Many signals are compressible by using some known transform coding scheme such as JPEG or JPEG-2000, which creates sparse representations in the transform basis. It is important to realize that a Mark Rothko color field painting is sparse, whereas a totally random screen test image is not. There is a great deal more information that cannot be compressed in the screen test image since there are significant and random pixel to pixel variations.

With this coarse understanding we can summarize the conventional paradigm for digital data acquisition: We uniformly sample data at the *Nyquist* rate and obtain N data points, which we then compress to K data points with K<<N using some threshold value of the data intensity and subsequently transmit or store this sparse data set. Using appropriate algorithms, we can decompress the reduced data sets with K data points back to N data points when we reconstruct the signal. This works quite well. There are well-known examples that show how close we get to reconstructing our originally sampled image after we omit 97.5 percent of all wavelet coefficients in the compression step.[3] However, as Mark Rothko and John Coltrane aficionados will tell you, there are differences in perception and quality when omitting wavelets coefficients with small values and subtle differences of hue and sounds can and will be noticed by the trained eye and ear. Humans can distinguish between 2.3 and 7.5 million colors (see Pointer and Attridge 1998: 52–4; Nickerson and Newhall 1943: 419–22), and about 340,000 tones (see Stevens and Davis 1938: 152–4).

Apart from desires to remain as close as possible to a uniformly sampled signal the question begs in particular for most commoditized images and sound recordings if instead of measuring lots of data and then compressing them, one could not attempt to measure only the data points which have significant wavelet coefficients above a pre-set threshold and omit the rest. The concept of compressed/compressive sampling/sensing[4] follows this approach and has proven that one *can* measure only a compressed data subset or something close to it and then find a way to reconstruct the signal N>>K. One is no longer measuring at or above the *Nyquist* rate but significantly below it since K<<N. This represents a radical departure from the traditional mode of measurement.

While on first sight it appears to be audacious, it has proven to be mathematically sound in the sense that it is highly probable given a sparse data set that one can reconstruct the image based on random subsets of data. These two pillars of compressed sensing, namely sparsity and random measurements, are crucial for the method to work. In compressed sampling one thus directly measures compressed data M<<N and thereby drastically reduces the amount of data needed to measure a signal *if* one finds a way to

reconstruct the total signal. More simplistically formulated: why measure the zero or below-the-threshold wavelet coefficients if "all" we have to do is measure the nonzero ones and then find a way to "add the zeroes back in." This appears very counterintuitive: how can one know what the compressed subset (the JPEG or MPEG) is before measuring it? Isn't the compression to a JPEG an analysis step one needs to do before one can omit data points with low or zero wavelet coefficients? How do I know which pixel is important and which is not? How do we find the "nonzero coefficients?"

Very complex mathematical details show that random measurements allow the acquisition of "compressed" data followed by a recovery relying on a technique called the *L1 minimization*. This procedure will provide us with the sparsest solution at a very high probability. The sparser the data are the higher the probability that we will recover the "uncompressed data"— without ever measuring them! The *L1 minimization* is a linear process and we have thus turned the conventional *Shannon-Nyquist* measurement philosophy on its head: we now measure random, nonlinear subsets of the data and recover the uncompressed data using a linear process (*L1 minimization*).

Of course this all hinges on the sparsity of our data since the mathematical core concept of compressed measurements is based on the realization that sparse signals can be recovered exactly with a high probability despite the fact that we are dealing with underdetermined linear systems of observations due to measuring below the *Nyquist* rate. Despite its counter-intuitive nature compressed sensing is based on sound and uncontested mathematics. While accepted in the sciences and engineering communities this new measurement mode will be difficult to explain to a lay audience and therefore be contested in the public arena, in particular when there are important consequences of such measurements. The early detection of cancer and other diseases based on measurements of only a subset of what can be measured will require justification and reassurances based on, as I will show below, nonfiscal cost functions such as radiation exposure or increased accuracy when imaging moving objects.

"Measure what can be measured and make measurable what cannot be measured" is a quote attributed to Galileo Galilei and has become the paradigm of the conventional *Shannon-Nyquist* measurement philosophy. Generations of scientists have and will continue to improve measurement devices such as telescopes, analogue and digital cameras and electron microscopes to "make measurable what cannot be measured." Minute signals indicating new phenomena are being confirmed after long periods of data acquisition and careful analysis. This mode of measurement will not be replaced by compressed sensing. Rather compressed sensing will complement the *Shannon-Nyquist* mode of measurement and allow new types of measurements previously not possible to be done.

"Measure what should be measured," a phrase coined by Strohmer (Strohmer 2012: 887–93) describes the operating philosophy of compressed sensing. In the conventional approach all data points are measured without taking into account the sparsity of the signal, whereas compressed sensing advocates to measure only the most important subset using random measurements in order to increase data flow, reduce storage requirements or required detector coverage. The need for compressed sensing and how it can be accomplished will be illustrated below using the paradigmatic example of the single pixel camera.

How Sparse, How Good?

The sparsity of signals allows the freedom to set threshold values in order to define what the cut-off intensity of the basis function is. This brings up the question of who needs or wants uniform sampling above the *Nyquist* limit and who will get to listen to a Coltrane saxophone solo recorded using compressed sensing or looks at a Mark Rothko painting with altered hue, or has his brain scanned less and faster? The important word in Strohmer's dictum is "should." We need to remind ourselves that compressed sensing works best with sparse data. In the case of imaging a screen test image which does not have a high degree of sparsity, a conventional measurement at the *Nyquist* frequency should be the choice of measurement mode. Many natural images are quite sparse and standard wavelet decomposition reveals often that most coefficients are actually very small. "Natural" images are highly structured and can be very efficiently represented using sparse representations.

Human perception is a very complex process with its own "biological" threshold values and in many cases the aspect of quality is difficult or even impossible to parameterize. The perception of color in a Mark Rothko painting serves as an example: the human eye's response to light spans from about 400 to 750 nm and varies with age. We have a standard for colorimetry that dates back to 1931 and was devised by the Commission Internationale de l'Eclairage (CIE), which assigns two coordinates, called CIE coordinates, to a particular color based on a photoluminescence measurement. In stark contrast, the standard method of characterizing the *quality* of light is to rely on a color rendering index which specifies how well a light source can illuminate, or render, the true color of an object. The color rendering index is determined by *comparing* the differences in the CIE color coordinates of an object illuminated first with a test source and then separately with a black body having the same correlated color temperature.[5]

This complex process reveals that the parameterization of quality often relies on comparison using human perception and can become a question of preference as in the case of the audiophile and art connoisseur, who resists the notion of a particular signal as being sparse. In the lighting industry lamps have different color rendering indices depending on if they are targeted for the U.S. or Asian market as the interaction with the skin complexion is an important quality factor. The U.S. initial response to fluorescent lighting was that it was "seen as cold" in comparison to the more energy inefficient incandescent light. There are thus external factors to a measurement that can play an important role in the decision as to which mode to use. These factors can often be best understood as external cost functions. In order to describe some of them and understand why they play such an important role in the implementation and use of compressed sensing I will now describe the single pixel camera, which is a paradigmatic compressed sensing device.

The Single Pixel Camera—a Compressed Sensing Device

An important aspect in the future design of sensing and measurement devices will be the hybridization of hardware and reconstruction algorithm as epitomized in the so-called single pixel camera developed by Richard Baraniuk from Rice University (see Duarte et al. 2008: 83–91).

The mathematical concept of random measurements needed for compressed sensing is incorporated in this new type of measurement device as follows: the light making up an image is focused from an array of mirrors and from there into a single detector. This arrangement resembles a projector running backwards. The mirror array consists of over 7000 little mirrors the size of a sand grain that can each be addressed in less than 10 microseconds

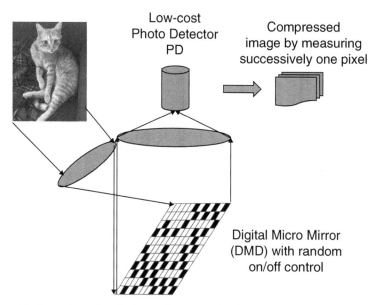

Figure 6.1 Operation of a single pixel camera: An image is projected onto a digital micro mirror (DMD), half of whose mirrors are switched randomly off. This image is then directed towards a second lens behind which a single photo detector (PD) registers the measurement. The mirrors are switched randomly and each time the PD measures one pixel. The single pixel camera thus creates a picture by measuring just one pixel over and over again which allows the recovery of the image.

Source: http://phys.org/news143210026.html.

either to reflect light into the detector or not. If one now randomly addresses groupings of half of the mirrors to reflect light into the mirror and the other half not and measures the intensity in the single detector a few hundred thousand times one is able to reconstruct the image from which the light comes. Each measurement contains information coming from half of the mirrors. A one-megapixel image would therefore require approximately 100,000 single pixel measurements representing only about 100 kilobytes of memory—a reduction of 90 percent in the number of measurements and with that measurement time, as well as storage space.

The "Cost Function" Driving the Use of Compressed Sensing

Compressed sensing has become *the* way to record data when an external "cost function" prevents or severely limits measurements in the classic *Nyquist-Shannon* mode. One way this "cost function" manifests itself is by the fact that certain detectors can cost upwards of a million dollar. To measure i.e. infrared images using compressed sensing one can thus use only a very small single pixel camera which enables measurements that would have prohibitive costs if measured in the *Nyquist-Shannon* mode where one would use a large detector to record as much signal as possible. Another way a cost function can promote this new type of measurement mode is due to the fact that in medical imaging one needs to minimize exposing patients to harmful ionizing radiation such as electrons and X-rays. Current estimates suggest that three to five percent of all new cancers diagnosed in the US are due to CAT scans (see Redberg and Smith-Bendmann 2014).

Drivers for this extensive use are the need to pay for these expensive instruments and the still prevalent philosophy of "measure what can be measured" endorsed by doctors and patients alike. Less exposure also reduces the data collection time and can thereby enable certain measurements: reducing the time patients need to hold their breath or remain still during an MRI scan. Compressed sensing will provide less perturbed images and is therefore already being used in pediatric medicine since it significantly reduces artefacts due to uncontrolled patient movement (see Vasanawala et al. 2010: 607–16). The amount of data reduction in MRI imaging by a factor of 2–10 comes with little or no impact of the quality of images (see Lustig, Donoho and Pauly 2007: 1182–95). In the case of brain imaging 3D images can be obtained in as little as 32 milliseconds (Hugger 2011: e28822).

Measuring only 10 percent or less of the signal will consume less energy, an issue important in space exploration and the deployment of autonomous sensor arrays for environmental or nuclear nonproliferation monitoring since it will increase the time period a probe can operate and send signals. In the short term many new applications in the design of active and autonomous sensing devices will drive the implementation of this new mode of measurement. Less energy use allows for significant further miniaturization and cost reductions. The "SeeChange" technology described in Dave Egger's novel

The Circle (2013) is technologically plausible, using compressed sensing as a measurement mode.

Challenges to Compressed Sensing

With these examples of cost functions in mind we can now discuss some challenges to the use of compressed sensing. Initially the counter-intuivite nature of compressed sensing can lead to a technically unfounded but very strong dismissal of this new mode of measurement. In the February 2010 edition of *Wired Magazine,* compressive sampling was introduced under the title *Fill in the Blanks: Using Math to Turn Lo-Res Datasets into Hi-Res Samples.*[6] Certain reader comments on the website display a lack of understanding of the general concept by assuming that common interpolation and not sophisticated and well established mathematical principles as outlined above are the basis of this method. This rejection is not so much a concern for the general public if one is imaging objects in intergalactic space or storing pictures of a birthday party. However, when using MRI which is now the premier diagnostic tool to detect cancer and vascular disease the "measure what can be measured" attitude is passionately evoked. Two quotes from the *Wired* website express this sentiment:

> Can you believe that someone would have such a fundamental misunderstanding of basic mathematics and information theory that they would base a medical diagnosis on features produced by data interpolation? I hope it's not my doctor doing it. Prettying up pictures is great. Looking for tumors, etc. is insane. By definition, you're looking for an aberration, which, by definition, this algorithm would not produce. (Iamorpa)

> Would you be willing to gamble your life on interpolated data where the shortcoming would be missing clinical pathology? A MRI image contains many subtle shades of gray in abstract shapes. For an artistic image clear, sharp edges and vivid colors may enhance, but to render a line on a diagnostic image that appears to abruptly end and restart may draw a complete artery where there is really an occlusion or to smooth out faint variations representing a brain tumor can be deadly. I'll pay for the long scan please (Grego).

It is important to recognize that compressed sensing can be rejected even by highly trained specialists who will not immediately take advantage of better or faster measurements available. As an example take a radiologist's unique skill set that allows him to scan huge amounts of data very fast and recognize and identify minute deviations. When varying the contrast in order to measure faster and expose the patient to less radiation such a highly trained specialist will no longer be able to identify anomalies and will insist on looking at the images with the same contrast settings they were used when he was trained—faster imaging and less radiation exposure is not always better when a significant amount of training and tacit knowledge to identify anomalous features relies on a specific contrast or image quality.

Clearly there will be early and late adopters of compressed sensing. The cost function be it money, radiation damage or speed with which one can measure will incentivize many to use this transformative imaging mode. In medical imaging using MRI, CAT or ultrasound devices this shift to taking advantage of compressed sensing is currently in progress. In the future this new mode of imaging will become the new standard, in particular when it is also used to train the tacit skills of medical specialists.

"Big Data Science"

As highlighted when describing the "single pixel camera" a significant reduction in measurements required to reconstruct an image by combining measurement and analysis modes will enable us to continue to cope with large data flows and storage requirements which do not increase commensurately with the data flow. Using compressed sensing we are able to measure a larger portion of the information content available by pushing back the limits of *Shannon-Nyquist* measurements.

In large-scale, complex scientific experiments which produce enormous amounts of data compressed sensing is already the mode of choice due to the need to "pre-select" events. The bottleneck in many measurements used to be at the detector level; now the commoditization of detectors which can be assembled into large arrays has moved the bottleneck to the processing/transmission/storage part of the measurement. The compact muon solenoid (CMS) detector at the Large Hadron Collider at CERN, the European nuclear science laboratory, produces 320 terabits per second of data.[7] Using a hardware-based triage, "triggers" select only about 800 gigabytes per second which will be characterized as "interesting events" and subsequently analyzed.

Owing to the appealing conceptual simplicity of compressed sensing's "second pillar," random measurement capabilities can directly be implemented in the design of the detector as described above for the single pixel camera. This hybridization of measurement and analysis is needed to be able to measure the gargantuan amounts of data without ending up with data flow bottlenecks and not enough storage space. Integrating sensing and data processing this way is "to bring mathematics into the lens" as Candes and Tao point out (2008: 14–7). An example highlighting the tremendous advances being made in data flow is the Sloan Digital Sky Survey[8], which started in the year 2000 and in its first few weeks collected more data than had been amassed during the entire history of astronomy. It will be surpassed by the Large Synoptic Survey Telescope[9] in Chile, which will measure 140 terabytes of data every five days.

Building on the advances and lessons learned in high energy and astronomy experiments compressed sensing will be employed in many other areas where the cost function is driven by the size of the already existing and growing data avalanche. The use of compressed sensing is growing rapidly and new devices are being designed and built. In analogue-to-digital (ADC) conversion technology a receiver/ADC chip, the "Random Modulator Pre-Integrator" (RMPI) has been built as one of the first electronic hardware devices based on the compressed sensing paradigm which will replace the conventional ADC in appropriate applications (see Becker 2011).

A consequence of this new mode of measurement is that data are no longer autonomous and complete because compressed sensing measurements rely on the premises of sparsity and measure only random subsets of all possible measurements. This will rekindle important discussions in the epistemology of measurement and change our view of what a discovery is. Discovery is no longer a "eureka moment" but a statistical war of attrition and the unease about increasingly theory-laden experiments will require a modification of what it means to discover something new. Serendipitous discoveries from very large data sets will in many cases be almost impossible since these particular events might be excluded from analysis precisely due to the fact that they do not exist within the current framework of knowledge at the time of measurement.

One such framework of knowledge is the Standard Model in Physics describing the occurrence and relationships of elementary particles. This model predicted the existence of the Higgs boson in order to explain the particular masses of elementary particles. Its discovery was based on looking for certain decay pattern allowed with the Standard Model of Physics and serves as a paradigm changing example of a theory-laden, large dataflow discovery. This type of measurement will become dominant in high energy physics and astronomy where data intensive experiments rely on a library of known and simulated processes to trigger further data analysis.

The continuous sampling of physical objects will amass enormous amounts of digital data as representations which can be re-analyzed at later stages using new algorithms and methodologies. We can learn something new using existing data and new algorithms. Antonioni's 1966 movie *Blow-Up* portrays the discovery of a body and a person with a gun after progressive enlargements of an existing picture. We need to be mindful, however, that there is a difference between having compressed data or used *Shannon-Nyquist* based measurements and simple experiments with Photoshop™ using raw and JPEG representations reveal striking differences when subsequently manipulating contrast or other image parameters. Software epistemology will need to be better understood as it can redefine what discovery means in such a context.

Compressed Sensing and Data-Driven Discovery—A New Mode of Inquiry

Stored data and information are now almost completely digitized: in 2013 the amount of globally stored information is estimated to be about 1,200 exabytes, of which 98 percent is in digital form. In 2007 the amount of data generated worldwide was larger than what we could store and transmit. In 2011 we measured twice as much data as we could store. The amount of data generated globally is growing by about 58 percent a year (see Gantz and Reinsel 2010).

At the same time a new strand of scientific inquiry, data-driven discovery, is establishing itself next to the traditional hypothesis-based mode. The measurement mode used to collect data is an important meta-property that needs to be taken into account when discussing results of such data-driven discoveries. There are reasons to suggest that this data-driven discovery mode will in the long run lead to fundamental changes in the daily practices and philosophical grounding of many sciences.

This type of "big data" analysis will be, at least initially based on the discovery of correlations which in many cases cannot be established unequivocally as causalities. A phenomenon identified by correlations might be a useful proxy for a highly probable prediction (i.e. flu tracking by *Google*) but shed no light on the reason for its existence or inner workings. A resurgence of devising heuristic rules and proxies in the physical sciences to correlate with desired physical (i.e. superconductivity, thermoelectricity) or chemical (i.e. catalytic) properties to explore vast parts of parameter space in chemical compound space are used in materials science ("materials genome initiative"). Simple proxies for properties are sought that can circumvent the traditional first-principles route using density functional theory or molecular dynamics in order to accelerate and enable new discoveries.

Filling Peterson's Quadrant—the Growth of Nonnomological Science

This increased use of heuristics might result in the emergence of new non-nomological research in various disciplines in particular the social sciences and economics. A convenient way to categorize types of scientific knowledge employs the scheme developed by Stokes (1997: 74) which assigns individual quadrants for pure basic (Bohr), use-inspired (Pasteur) and pure applied (Edison) research. The first quadrant contains research with little considerations for use and fundamental understanding. As Stokes writes: "This quadrant includes research that systematically explores particular phenomena without having in view either general explanatory objectives or any applied use to which the results will be put ... " (ibid.).

Stokes named this quadrant Peterson's quadrant to point out that Peterson's *Guide to the Birds of North America* exemplifies such systematic research.

Within the framework of this classification scheme one can describe dynamic pathways in which research progresses from one quadrant to the next. The progression from Bohr's over Pasteur's to Edison's quadrant can be used in retro perspective to describe various phases in the evolution of modern electronics from quantum mechanics to the modern computer chip. An important transition from Peterson's to Bohr's quadrant is typified by Charles Darwin's theory of evolution described in *The Origin of Species*. Darwin relied on a vast

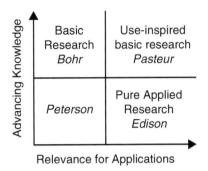

Figure 6.2 Stokes's Quadrants.
Source: own image based on D. E. Stokes, *Pasteur's Quadrant: Basic Science and Technological Innovation* (Washington: Brookings Institute Press, 1997).

amount of not understood correlations as well as many disparate observations. These collations of observations in Peterson's quadrant can thus lead to important pre-cursors of theories, proxies and correlations that can find applications in pure applied research (Edison's quadrant).

Another important role of this quadrant as pointed out by Stokes is education and the development of skills. The availability of large amounts of digitized data triggered the development of *Google*'s page ranking algorithm, unstructured data bases and many other problems needed to be solved to correlate observations using large heterogeneous data sets. These methodological problems are in themselves worth pursuing as they can sharpen our tools to "dig deeper" into certain types of data. In analogy to the above-mentioned scenario from the film *Blow-Up* the existence of large data bases will allow us to develop new algorithms and methodologies that could lead to the discovery of new features and correlations in already existing data and allow their benchmarking.

Subsequently, a complementary hypothesis-driven inquiry mode might be able to devise measurements to verify whether the heuristic rules and proxies developed in Peterson's quadrant are useful and can be reduced to causal chains. While large data correlations might point towards some new, unforeseen and unexpected hypotheses many such correlations are likely not to be reducible to discipline-specific models and causation chains. Nevertheless, these correlations might still be useful as rules and become test cases for meta-theories used for instance in complexity science where strong causality is often suspended in favour of a more heuristic and empirical view of the sciences.

Some correlations might be used to develop "toy models" rather than complex and very detailed models with too many parameters. An example for a "toy model" theory is the theory of "self-organized criticality" developed by Per Bak (1996). The power of these oversimplified models lies not in the prediction of individual events but in the description of complex systems such as financial markets, forest fires, earthquakes and the size of avalanches in a sand pile. These types of theories might undergo a renaissance as our data bases grow. As Peter Norvig writes: "Simple models and a lot of data trump more elaborate models based on less data"' (2009: 8–12).

Understanding the world without relying on hypotheses will become a new type of exploration based on probabilities and correlations. Noncausal analyses will help us see the world within a context of "what" and not "why." While we have an intuitive desire for causal connections many fields of exploration in particular in the "soft" and idiographic sciences will be driven by the discovery of correlations. It is too early to assess how the shift from causation to correlation will change the operational pragmatism in different scientific fields.

In conclusion, compressive sensing allows us to keep up with the in the foreseeable future continuing massive flow of data and enables, as a complementary tool the measurement of sparse data when an external cost function (i.e. price, speed, stability) makes measurement, transmission, and data storage in the classical *Shannon-Nyquist* mode no longer feasible. It will therefore make a significant contribution to our digital universe and provide ample research data for Peterson's quadrant which can fuel Bohr's, Pasteur's or Edison's quadrant by classical hypothesis-driven inquiries or use heuristic rules and proxies to enable pure applied research where correlation can be enough to predict the behavior of large complex systems without an underlying causal model.

Notes

1 See http://en.wikipedia.org/wiki/Balance_puzzle.
2 Another approximation often used to approximate and create images, particular mountain- and other landscapes is based on self-affine fractals. This is used in animation and virtual reality images (see Voss 1988).
3 See Figure1 in http://www.ams.org/samplings/math-history/hap7-pixel.pdf.
4 The terms *compressive sensing*, *compressed sensing* or *sampling* are used interchangeable in the scientific literature.
5 A light source's color temperature is the temperature an ideal black body would have that is closest in hue to that of the light source in question. Diffuse sunlight has a color temperature between 5,700 and 6,500 degrees Kelvin, whereas a candle's color temperature is near 2,000 degrees Kelvin.
6 See http://www.wired.com/magazine/2010/02/ff_algorithm/.
7 See http://cms.web.cern.ch/.
8 See http://www.sdss.org/.
9 See http://www.lsst.org/lsst/.

References

Bak, P. 1996. *How Nature Works: The Science of Self-Organized Criticality*. New York: Copernicus.

Becker, S. R. 2011. *Practical Compressed Sensing: Modern Data Acquisition and Signal Processing*. PhD dissertation, California Institute of Technology, Pasadena, California.

Candes, E. J. and Tao, T. 2008. Reflections on Compressed Sensing. *IEEE Information Theory Society Newsletter* (December): 14–7.

Duarte, M., Davenport, M., and Takhar, D. et al. 2008. Single-pixel Imaging via Compressive Sampling. *IEEE Signal Processing Magazine* 25:2: 83–91.

Eggert, D. 2013. *The Circle*. New York: Knopf.

Gantz, J. and Reinsel, D. 2010. The Digital Universe Decade—Are You Ready? *IDC White Paper* (May).

Halevy, A., Norvig, P., and Pereira, F. 2009. The Unreasonable Effectiveness of Data. *IEEE Intelligent Systems* 24:2: 8–12.

Hugger T. et al. 2011. Fast Undersampled Functional Magnetic Resonance Imaging Using Nonlinear Regularized Parallel Image Reconstruction. *PloS One* 6: e28822.

Lustig, M., Donoho D., and Pauly, J. M. 2007. Sparse MRI: The Application of Compressed Sensing for Rapid MR Imaging. *Magnetic Resonance in Medicine* 58: 1182–95.

Nickerson, D. and Newhall, S. M. 1943. Number of Discernible Object Colors Is a Conundrum. *Journal of the Optical Society of America* 33: 419–22.

Nyquist, H. 1928. Certain Topics in Telegraph Transmission Theory. *Trans. AIEE* 47: 617–44.

Pointer, M. R. and Attridge, G. G. 1998. The Number of Discernible Colors. *Color Research & Application* 23: 52–4.

Redberg, R. F. and Smith-Bendmann, R. 2014. We Are Giving Ourselves Cancer. *New York Times* Opinion January 30, https://www.nytimes.com/2014/01/31/opinion/we-are-giving-ourselves-cancer.html, accessed December 15, 2016.

Shannon, C. E. 1949. Communication in the Presence of Noise. *Proceedings of the IRE* 37: 10–21.

Stevens, S. S. and Davis, H. 1938. *Hearing. Its Psychology and Physiology*. New York: Wiley.

Stokes, D. E. 1997. *Pasteur's Quadrant: Basic Science and Technological Innovation*. Washington: Brookings Institute Press.

Strohmer, T. 2012. Measure What Should be Measured: Progress and Challenges in Compressive Sensing. *IEEE Signal Processing Letters* 19:12: 887–93.

Vasanawala, S. S. et al. 2010. Improved Pediatric MR Imaging with Compressed Sensing. *Radiology* 256: 607–16.

Voss, R. F. 1988. Fractals in Nature: from Characterization to Simulation. In: *The Science of Fractal Images*, edited by H.-O. Peitgen and D. Saupe, 21–70, New York: Springer.

Internet Sources:

http://cms.web.cern.ch/, accessed March 11, 2016.

http://en.wikipedia.org/wiki/Balance_puzzle, accessed March 11, 2016.

http://www.ams.org/samplings/math-history/hap7-pixel.pdf, accessed March 11, 2016.

http://www.lsst.org/lsst/, accessed March 11, 2016.

http://www.sdss.org/, accessed March 11, 2016.

http://www.wired.com/magazine/2010/02/ff_algorithm/, accessed March 11, 2016.

7 The Altered Image: Composite Figures and Evidential Reasoning With Mechanically Produced Images

Laura Perini

Introduction

Visual displays produced by mechanized imaging techniques—in which the visible features of the display are the result of the interaction between specimen and imaging technology—play important roles in generating scientific knowledge. Mechanically produced images (MPIs) include photographs, X-ray films, and autoradiographs.[1] MPIs are often used by scientists to evaluate hypotheses. They typically play that evidential role by functioning as visual representations, defined as visual displays in which some spatial features of the display are interpreted to represent features of the referent (see Perini 2005). However, research talks and publications rarely include figures that are simply reproductions of the MPIs originally produced; instead the MPIs are altered in various ways. This is somewhat surprising, since the mechanistic connection between image and the specimen seems to be crucial to the evidential roles MPIs play. This contribution aims to explore the epistemic issues involved with the use of altered MPIs in scientific reasoning.

It is easy to overlook the fact that there is an important step taken between producing an MPI, and using that image as a visual representation. However, input from two quite different sources shows that the representational role that an MPI plays depends on factors beyond the image and its mechanistic link to the specimen—and that those factors matter for understanding how MPIs function as evidence.

First, philosophical analysis of depiction has shown that two-dimensional visual arrays do not intrinsically depict—or in any way represent—anything at all.[2] Furthermore, relations between the image and what is depicted, such as similarity or causal connections, do not suffice to determine the content of a visual array, or even that it is a visual representation. As Goodman (1976) argued, you can use an image to represent just about anything you want, and you could use it for nonrepresentational purposes. The recent discussion of scientific representation has absorbed this point; for example, Giere (2006) and van Fraassen (2008) both stress the pragmatic nature of scientific representation in general, and apply this to images in particular. The upshot for the use of MPIs in science is that they do not intrinsically represent anything relevant to assessing a hypothesis or theory; they *could* be used to represent just about anything.

There is a second body of work, by historians and sociologists of science, showing that there is an important gap between an image and its use as a visual representation. Those studies demonstrate that scientists using imaging technologies in their research often produce images that they cannot initially use as representations—in particular, they do not know how to use them as representations that are relevant to assessing a particular hypothesis, or enabling them to ask a further question about the subject matter they are studying.

The ethnographic studies show that interpreting an MPI in order to use it as a representation can involve intensively social and temporally extended work by the very experts who produced the image, and that the interpretive work to assign representational content to an MPI can be highly dependent on local resources and interactions among research group members. If the evidential support an MPI provides depends on its representational content, but that is assigned to the image by interpreting it in a way that depends on factors that are not available to the wider scientific community, then the researchers face a communication problem with a distinctive epistemic dimension. I will begin by clarifying the interpretive issues, and then discuss the representational tactics scientists use to solve the communication problem.

From Image to Representation: the Interpretive Struggle

Scientists using imaging technologies sometimes produce images that they are initially unable to use to represent the kinds of things that matter for their research. Of course, some MPIs present little difficulty of this kind; scientists can readily use them to represent items of relevance to their research. We are all familiar with imaging technologies adapted to diagnostic purposes in medicine. Doctors usually have little difficulty using an X-ray film of a human torso to represent standard anatomical features such as ribs. Sometimes, scientists will conduct research in order to learn how visible features in the MPIs correlate with known properties of imaged subjects.[3] In other cases, imaging technologies are refined for use on specific kinds of research subjects, to facilitate this ease of interpretation.[4]

Although some MPIs can be easily used as visual representations, sometimes researchers struggle to use the MPI as a visual representation—to use their visual experience of the image to comprehend its visible features as representing features of the specimen. This sort of case was brought to light in *The Fixation of (Visual) Evidence* (Amann and Knorr Cetina 1988). Their subject was a molecular genetics lab studying the first step in gene expression, the initiation of transcription. The lab produced data in the form of autoradiographs of electrophoresis gels. Amann and Knorr Cetina do not provide details about the experiments, but a common experimental approach involves mixing various biological components (proteins, DNA, nucleotides, etc.) in order to determine what factors affect the initiation of transcription, and how they do so. Experiments can be designed so that a hypothesis—for example a claim that a particular protein functions as a repressor on a certain gene—can be evaluated based on lengths of nucleic acid molecules produced by a particular experimental mix of components. As the nucleotide fragments produced in these reactions are too small to see, further work must be done in order to determine what happened in a particular experiment.

The first step is gel electrophoresis, a technique that separates charged macromolecules by size. In gel electrophoresis, samples are loaded into "wells" at the top of the gel, and an electric current pushes charged material (like nucleic acids) down the gel. Small molecules make their way through the gel matrix faster than larger ones, so the molecules are separated by size, with larger ones closer to the top of the gel and smaller ones farther below. The array of macromolecules in the gel is then visualized by autoradiography. Autoradiographs are images made by exposing X-ray film to radioactive samples; radiation produces dark areas on the film when it is developed. An autoradiograph of a gel electrophoresis experiment will usually have several columns in which some dark bands are visible. Each column corresponds to the material loaded into a well in the gel, and the bands in a column correspond to charged material of a particular size.

In an experiment on gene expression, radioactive DNA or RNA is used, and gel electrophoresis separates nucleic acid fragments of different lengths. Researchers usually load at least one well with markers of known lengths. When the MPI is produced, the researchers can see columns of bands, and they can use the MPI as a representation of certain features of the gel—the location of radioactive nucleic acid in the gel—by interpreting the spatial array of the bands in the image as representing the spatial array of nucleic acid in the gel. Furthermore, they can interpret the MPI as representing larger fragments at the top of the image and smaller ones in the bottom, by interpreting vertical spatial relations as relations in nucleic acid fragment size.

It would seem to be easy, then, to use an autoradiograph to represent specific nucleic acid lengths. To interpret individual bands in the marker lane as representing the specific lengths the marker mix was supposed to contain, apply knowledge about the length of the fragments in the marker mixture, namely in a way that the larger sized fragments are assigned to the upper bands, and smaller fragments to the lower bands—using the interpretation of vertical spatial relations in the MPI as representing gradation in molecule size. Then the interpretation of horizontal spatial relations—different columns representing different experimental mixtures—can be used to interpret bands in experimental lanes: look horizontally from a marker band to an experimental lane, and try to find a band in that lane that is a similar vertical distance from the "well" area at the top, in order to assign a specific meaning (in terms of fragment length) to that band in the MPI.

This sounds easy, and sometimes it is. However, as Amann and Knorr Cetina show, sometimes it is very difficult. Sometimes scientists at first are *not* able to interpret their autoradiograph in this way, and will not use it to represent specific nucleic acid fragment lengths, the thing that they need to do in order for the autoradiograph to be used to evaluate a hypothesis about whether or not gene expression was initiated.

Amann and Knorr Cetina analyze the processes that the researchers use in order to interpret the autoradiograph as a representation. The researchers gather multiple times in pairs and small groups, offering, and critiquing, different suggestions for interpreting the autoradiograph. The scientists visually inspect the film while talking about the details of that experiment; they retrieve other autoradiographs produced by the lab for comparison; they discuss research publications from other sources; and they challenge each other's proposals about what specific parts of the autoradiograph should be taken as representing. Above, I mentioned the philosophical issue connected to the use of MPIs as visual representations: Goodman's point that just about anything can be used to represent just about anything (see Goodman 1976). Goodman's philosophical analysis implies that there is a sense in which these researchers *could* have used the autoradiograph to represent something of relevance to their hypothesis, without all the effort. They *could* have just stipulated that certain bands represent certain nucleic acid fragments. From the transcripts Amann and Knorr Cetina provide, this option was not even considered. Amann and Knorr Cetina describe scientists in a lab pooling knowledge, critical abilities, and additional images in order collectively to achieve an interpretation of the MPI as a representation of specific kinds of features that are relevant to evaluating a hypothesis. The scientists in this case seem to be unable to use the autoradiograph as a representation in light of adherence to epistemic norms.

Amann and Knorr Cetina (1988) argue that one prominent feature of the collective effort to interpret the image is the use of oppositional questioning. The ability to stand up to oppositional questioning includes probing for sources of error, such as mistakes during the experimental procedure. The study also shows that in the cases documented, coherence with other sources of information also plays an important role in evaluating an interpretation (see ibid.). The norm of coherence is applied to proposed uses of the image used as a visual representation, in light of beliefs about the experiment, relevant scientific models, and visual representations including other MPIs whose representational content has been settled. Those norms, along with the resources assembled in the discussion process, and the visual appearance of the MPI itself, constrain scientists' representational use of MPIs considerably—and sometimes make it difficult for scientists to use an MPI in a way that meets those constraints.

The case described above shows that, faced with an autoradiograph produced from experimental samples with various mixtures of nucleic and proteins that regulate DNA expression, adherence to epistemic norms prevents the researchers from initially using the image to represent the specific things that matter for evaluating a hypothesis of interest to them. When the scientists reach the point where they can use the autoradiograph as a visual representation of specific nucleic acid fragments, they have reached this point as a result of an intensively social process that made use of local resources of knowledge, experimental skills, and other autoradiographs.

This puts the researchers in a difficult position: scientific knowledge is a public matter, requiring evidence that is comprehensible to a community of researchers. Scientists publishing a research report that uses a difficult-to-interpret MPI as support for a hypothesis cannot expect their audience to be able to comprehend that image as a representation with the same content that they now do. The audience does not have the resources available to the originating lab, and without those resources, the norms of coherence and ability to withstand oppositional questioning offer little constraint on image interpretation. That in turn suggests that the constraints are serving a further value: coherence and the ability of an interpretation to withstand challenges are pursued because what the scientists want out of the MPI, used as a visual representation, is accuracy.

What scientists do *not* typically do is publish supplementary materials so that the audience can conduct a similar interpretive process of the original MPI as its producers did. Instead, what they present to the public as their reasoning about imaging experiments are figures that have been manipulated by the researchers in significant ways. One important tactic involves altering the visible form of an MPI—sometimes dramatically—and publishing an altered image that can be more readily and reliably interpreted as a representation of the features that matter for the study than could the original MPI. Another common approach to make use of other kinds of visual representations is juxtaposing MPIs with diagrams.

Deletion and Rearrangement of MPIs

Consider Figure 7.1, which was published in a study of the circadian cycle of monarch butterflies. The study focuses on two monarch genes, *cry* 1 and *cry* 2, which are similar to two different known genes (*Drosophila cry*, and the vertebrate *cry* gene, respectively). The goal of the study was to investigate how each of the two monarch *cry* genes functions in the monarch circadian cycle, which involves changes in the amounts and activities of several proteins during day-night cycles. The scientists conducted a set of experiments in which double-stranded RNA targeting the *cry* 1 gene is added to cells cultured in a light-dark cycle, at the onset of a light period. This treatment should inhibit *cry* 1 gene expression. A double-stranded RNA inhibiting the noncircadian gene *GFP* was used as a control. The goal of the experiment was to determine the effect of inhibiting *cry* 1 expression on the levels of various proteins associated with circadian cycles.

Samples are taken at intervals after the onset of the light period. In order to visualize the proteins, samples are separated by gel electrophoresis. Next, the gels were treated with solutions that target a specific protein in the gel (CRY 1, TIM, PER, or CRY 2), and attach a substance that has the capacity to expose film. In each of the panels A-D, the left side image is divided horizontally, with one band labeled in each of the two halves—for example, the CRY 1 protein band is labeled in A. The authors are not asking the audience to decide which band corresponds to the protein that was assayed; the labels indicate that particular bands are to be taken as representing specific proteins. The original gel images have been cut and rearranged so that the columns for CRY 1 protein from the "ds *cry* 1" condition are placed directly under those from the control condition, labeled "ds *GFP*." If the samples from both control and experimental reaction were separated on the same gel, the CRY 1 protein would be at about the same horizontal position under both conditions.

The array presented here must either be an image of a single gel that has been cut and rearranged, or images from two different gels that have been cropped and positioned one above the other (for each of A through D). In the published panels, the columns for the control are positioned directly above those for the experimental condition, matching the time points of the samples. This arrangement makes it easy to compare changes in size and darkness across the row of bands in the control with changes across the row when *cry* 1 is inhibited.

I also noted above that the interpretation of MPIs as visual representations seems to be guided by norms. Another norm involved in the use of MPIs for representational purposes is that of objectivity. Daston and Galison (1992) study the history of objectivity through an examination of a particular type of scientific publication whose core is a collection of figures: the atlas. They chart the rise in prominence of an epistemic ideal calling for minimizing intervention and interpretation on the part of the researcher—"mechanical objectivity." MPIs offer a way for scientists to approach the ideal of mechanical objectivity, since the mechanized interaction between imaging technology and the specimen produce an array that is dependent on that causal interaction, and less on choices made by the researcher, compared to the options available with a hand-produced rendering.

What Daston and Galison are focused on, however, is not how the figures published in atlases function as representations of particular specimens, but rather, how those representations of particular individuals—real or idealized—in turn represent a class of individuals. There is no indication that the MPIs they discuss present any difficulties in interpreting those images as representations of specimen features—the kind of difficulties that Amann and Knorr Cetina describe. The atlas authors that Daston and Galison present as examples of scientists hewing to the ideal of mechanical objectivity are abstaining from theorizing about the class of individuals; they present their audience MPIs that function as visual representations of specimen features, and place the cognitive task of drawing conclusions about the class from the individuals represented largely on the shoulders of their readers.

In the kind of experimental work that produces hard-to-interpret MPIs, the scientists are in a different situation: they are not starting with images that are readily interpreted as visual representations of the kind of specimen features that matter for evaluating experimental results. Their approach to publishing those MPIs contrasts starkly with that of atlas authors aiming for mechanical objectivity. As seen in Figure 7.1, published gel images are labeled (so the scientists are not refraining from interpreting the MPI), as well as altered. This is very common in published MPIs, and typical for gel images in particular. They are almost always cropped; often parts of an image are cut out and juxtaposed, and sometimes parts of *different* gel images are juxtaposed, to put the two into visual relations that in turn represent particular items as standing in relations that bear on the hypothesis being evaluated. However, attention to what in the original image is relevant to the researcher's aims—what they are trying to learn, and what they need to communicate—shows that the reduction in mechanical objectivity is selective, and subordinate to the researchers' earlier conclusions about how to understand the MPI as a visual representation.

In order to evaluate hypotheses about how cry 1 affects the circadian cycle, the authors must use particular bands in those MPIs to represent particular proteins, across the different columns taken to represent successive time points. The authors provide an interpretation for the audience, and alter the images in ways to guide comprehension of the important relationships among the data. The experiment is testing the effect of inhibiting the cry 1 gene on the level of CRY 1 protein (A) and three other proteins involved in circadian cycles (B–D). The image is arranged so that experimental and control bands corresponding to samples from the same time points are in vertical columns. This allows for comparison of how the amount of that protein changes throughout the experiment, when cry 1 is inhibited vs. the control. Although the interventions in crafting the figure are significant, one key aspect of the original MPI has (presumably) not been altered: the darkness of the bands now used to represent particular proteins. The preservation of the range of darkness of the bands in a horizontal row is crucial. The fact that the relative darkness of the bands is a result of the interaction between particular specimens and the imaging technology, and not something drawn in by the researchers, is essential in this case for the support the figure provides for their conclusion.

Although in general the authors are not adhering to a norm of nonintervention, they are selective, deleting some information the scientists think they do not need, while preserving some other information they think is embodied in the MPI. The rearrangement reflects the scientists' interpretation of the original MPIs (e.g. with particular bands representing particular proteins), but facilitates comprehension by a wider audience: the parts of the gel images conveying rows of bands are set in spatial relationships with each other that more effectively communicate about the relationship between the control and experimental condition. The ideal of mechanical objectivity has been largely, although not entirely, sacrificed in the service of communicating evidentially relevant content.

Diagrammatic Rendering

Figure 7.1 exemplifies another tactic scientists use when communicating with MPIs: visual juxtaposition of diagrammatic visual representations together with MPIs.[5] The two types of image are often paired in order to support a hypothesis: MPIs have a visible form caused by the interaction between specimen and imaging technology, and are used to represent specimen features— often in detail. Diagrams convey relatively abstract content and are often

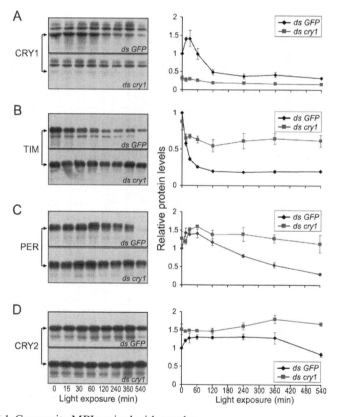

Figure 7.1 Composite MPIs paired with graphs.

Source: H. Zhu Haisun and I. Sauman, Q. Yuan, A. Casselman, M. Emery-Le, P. Emery, and S. Reppert, "Cryptochromes define a novel circadian clock mechanism in Monarch butterflies that may underlie sun compass navigation," *PLoS Biology*, 6:1 (2008), Doi: 10.1371/journal.pbio.0060004.

designed to make the relevance of that content to the hypothesis at issue salient.

Figure 7.1 provides an example of this tactic; the MPIs on the left are paired with graphs on the right. The graphs are plots of data derived from analyzing the MPIs: band intensities are measured, and the mean band intensity (data from three experiments) is plotted in a graph. There is a distinctive difference in the visual appearance of the MPIs vs. the graphs; there are also differences in their content, when these images are construed as visual representations. The MPIs are used to represent the location and amounts of proteins in a gel by comprehending the dark bands as representing particular proteins, so the specific darkness, size, and shape of the labeled bands all matter. For a graph on the right, on the other hand, only a few visual aspects of the image contribute to its content: the specific locations of data symbols and lines matter, but the width of the lines and the size of the data symbols do not. A color difference (black vs. red in the original, reprinted in greyscale) is used to refer to different conditions, but the specific shade does not matter.

What is the value of presenting the graph along with the MPI? The authors have labeled the bands in the MPIs, assigning them referents. However, the MPIs in this case need to represent not just the presence of specific proteins in a particular gel lane, but the relative amount of that protein over the time of light exposure. The graphs provide three advantages: first, unlike the MPIs, the graphs convey little content besides information that matters for supporting the hypothesis; second, the graphs convey the precise details needed to evaluate the hypothesis more reliably than can the MPIs; and third, the graphs are formatted to represent more prominently the relations among aspects of the data that have evidential relevance to the hypothesis.

First, the MPIs in Figure 7.1 are used to represent the spatial location and amounts of proteins in the gels, through the spatial location of the bands in the images, and the darkness of those bands, respectively. With the addition of labels, they represent the specific location of particular protein bands, in addition to the location of other material in the gel (not labeled in the figure). In the graphs, symbols (diamond vs. rectangle) and color differences (black vs. red in the original figure) are used to indicate that a value is associated with a particular protein. Spatial locations in the graphs are used to represent the relation between the optical intensity of particular bands in the MPIs to particular time points. In the MPI, spatial features of the image are used to represent spatially distributed features of the experimental artefact; in the graph, spatial relations are not used to represent spatial relations, but relations among other kinds of properties. There is another difference in content between the two; the MPI conveys details about the gel that is not needed for assessing a hypothesis about how the *cry* genes are involved in the circadian cycle—information that is not included in the graphs.

Second, and even more importantly for using the MPI as support for the researchers' conclusions, however, information that *is* relevant to the researcher's concerns is not readily accessible by visual inspection of the MPI.

What matters for drawing conclusions about the role of *cry* 1 is the change in amount of protein after the onset of light. Limitations of human visual capacities make extracting the specific information needed from the MPI difficult. Consider the MPI in Figure 7.1, part A. The darkness of the CRY 1 protein bands in the first four lanes looks about equal, and then there is a visually distinct drop in the next lanes. However, the graph plots an initial increase, then a drop within the first four lanes. While this graph compiles results from two additional gels, the difference between the relative amounts of protein people comprehend the MPI to represent (in the absence of the diagram) vs. what the diagram conveys cannot be attributed simply to the addition of results from the other two gels. Films register differences that human vision cannot reliably detect. For this reason, the film itself is measured and analyzed, in order to extract the information embodied in the MPI; subsequently, the results of that analysis are communicated in a diagrammatic visual representation.[6]

The importance of this is seen in Figure 7.1, parts C and D. The graphs represent clearly two different patterns of decline in the amount of PER and CRY2 compared to their respective controls; these patterns are not visible in the MPIs. In this case, failure to analyze the original MPI, and reliance on using the original MPIs as visual representations of the amount of proteins would have led researchers to draw different conclusions about the interactions among these proteins, due to inability to make the needed visual discriminations. Thus one benefit of the graph is that it provides a way to represent information about the experimental results with the precision needed to reliably use those results to evaluate hypotheses about the role of *cry* 1.

Finally, the graph plots values for both the control condition and *cry* 1 inhibited condition into the same space, while providing visual cues so that the reader can easily interpret particular data points as representing a particular condition. This contrasts with the MPI, in which bands for one condition are positioned above, but separate from, the other. Plotting values in the same space provides a visually prominent representation of the relation between values for control and experimental conditions over time; the vertical distance between the control vs. experimental value is easy to see, and because vertical position is interpreted according to the label on the y-axis of the graph, the graph allows the reader to determine the difference between the quantities with some precision. The graph conveys key relations among data points—for example that ds *cry* 1 reduces the amount that the PER bands fade during the light period, compared to the control condition—that are directly relevant to establishing the role of *cry* 1.

Conclusion

While the philosophical literature on scientific images is growing, some of the most distinctive features of images published in contemporary research reports, such as alteration of MPIs and combinations of different kinds of images in a single figure, have had little attention. This exploration provides a first step in that direction. The backdrop for this first step is the literature on images that shows both that the products of scientific imaging techniques required interpretation in order to function as visual representations, and that laboratory scientists impose significant constraints on what an individual MPI should be used to represent.

Analysis of published MPIs demonstrates the priority of formulating that interpretation: it guides the sorts of alterations to the MPI that the scientists are (or are not) willing to make. That initial struggle to interpret an MPI as a visual representation also has a hypothesis-oriented component: the difficulty arises in being able to use the image as a representation of the specific features relevant to evaluating hypotheses about the experimental subject matter. The complicated design of contemporary figures in research publications reflects these key concerns, which drive decisions about alterations of MPIs, about measuring and analyzing the MPIs and presenting additional visual representations to communicate those results, as needed in order to communicate a strong case for the conclusions the authors defend in the publication.

Notes

1 Some imaging technologies, like Magnetic Resonance Imaging (MRI) and Positron Emission Tomography (PET) involve a mechanized interaction between a specimen, from which data is collected in digital format, and then subjected to extensive mathematical analysis. It is the output of the analysis that is ultimately visualized. This contribution will focus on more direct imaging techniques, in which the initial output of the interaction between specimen and imaging system is a visual display, with little or no intervening analysis.
2 Depictions are a subset of visual representations as I have defined them here. For more on depiction vs. nondepictive types of visual representation in science see Perini (2012).
3 On X-ray imaging and human anatomy see Pasveer (2006).
4 On the development of magnetic resonance imaging see Semczyszcyn (2010).
5 The differences between the MPIs and the graphs they are paired with in this example will be discussed below; for a more comprehensive discussion of the difference between diagrams and nondiagrammatic visual representations see Perini (2013).
6 The analysis also includes normalizing the signal against that for a noncircadian protein.

References

Amann, K. and Knorr Cetina, K. 1988. The Fixation of Visual Evidence. *Human Studies* 11:2–3: 133–69.
Daston and P. Galison, L. 1992. The Image of Objectivity. *Representations* 40: 81–128.
Giere, R. 2006. *Scientific Perspectivism*. Chicago: University of Chicago Press.
Goodman N. 1976. *Languages of Art: An Approach to the Theory of Symbols*. Indianapolis: Hackett.

Pasveer, B. 2006. Representing or Mediating: A History and Philosophy of X-ray Images in Medicine. In: Pauwels L. (ed.), In: *Visual Cultures of Science: Rethinking Representational Practices in Knowledge Building and Science Communication*, edited by L. Pauwels, 41–62, Hanover, NH: Dartmouth College Press.

Perini, L. 2013. Diagrams in Biology. *The Knowledge Engineering Review* 28:*3*: 273–86.

Perini, L. 2012. Depiction, Detection, and the Epistemic Value of Photography. *The Journal of Aesthetics and Art Criticism* 70:*1*: 151–60.

Perini, L. 2005. Visual Representations and their Role in Confirmation. *Philosophy of Science* 72:*5*: 913–26.

Semczyszcyn, N. 2010. *Signal into Vision: Medical Imaging as Instrumentally Aided Perception*. PhD dissertation, University of British Columbia.

Van Fraassen, B. 2008. *Scientific Representation: Paradoxes of Perspective*. New York: Oxford University Press.

Zhu, H., Sauman, I., Yuan, Q., Casselman, A., Emery-Le, M., Emery, P., and Reppert, S. 2008. Cryptochromes Define a Novel Circadian Clock Mechanism in Monarch Butterflies That May Underlie Sun Compass Navigation. *PLoS Biology* 6(*1*):e4, Doi: 10.1371/journal.pbio.0060004.

8   Visual Data—Reasons to be Relied on?

Nicola Mößner

Introduction

In modern science, the output of measurement processes is often visual representations of the data detected—*visual data* for short.[1] Predominately, these images have instruments as their sources which are often combined with information technology devices processing and displaying the data measured. For the sake of simplicity, I take (digital) photographs as my main example here, although I believe that the crucial points can also be applied to other kinds of visualizations.

In regarding photography as an instance of measurement devices, I follow Patrick Maynard's thesis (see Maynard 2000: 5) that *photography* should be understood as a *family of technologies*.[2] It denotes a certain process of marking (specially prepared) surfaces with the aid of rays of light or other kinds of radiation (see ibid.: 5, 9). Taken in this more general sense, it appears unproblematic to state that photography not only allows *depiction* as it is found in family snapshots and the like, but serves different functions that are based on the above-mentioned process. Accordingly, Maynard points out that photography may also be used for *detective* purposes (see ibid.: 120 ff.).

Furthermore, it is also possible that detection and depiction are both put to use in a measurement process, thus resulting in a visual representation of the research object. An example of this would be to use photometric methods to determine the magnitude and age of a star—say *Antares*—by measuring its color. The whole process consists of different steps of measurement, calculation and inferences[3]—but let us concentrate on measurement alone. An important aspect then is to use different color filters so that only an amount of light per defined range of wavelength (commonly known as "color") can be detected by the camera. Attaching a blue filter to our camera means that a hot blue star will appear brighter on our photograph than a cooler red one. A photograph of *Antares* taken with the aid of such filters can then be called a measurement result. This example shows that in the realm of detection and discovery scientific reasoning often essentially involves visual data.

Normally, relevant measurement results are also published to make them available to the scientific community. In this context, Laura Perini (2005) has pointed out that visualizations should not be regarded as mere illustrations, but can play an essential role, i.e. make significant epistemic contributions to the linguistically presented argument. Particularly with regard to this second usage of scientific visualizations, however, Martin Kemp (2012) arouses our suspicions about their apparent and often cited *evidential status*. He raises the question why we should believe that there is such a strange creature as a mammal with a beak[4] just because its picture was published in a journal. Even though his example is about handmade drawings, we can apply his point here, too. What justifies a scientist in reasoning with the aid of (published) visualizations, taking them as measurement results? Believing without reasons is epistemically irresponsible—it is *irrational*. Accordingly, in this contribution we will analyze whether there are any reasons to support our epistemic practices concerning scientific visualizations in the above-mentioned senses and whether these are good reasons for us to rely on.

Kinds and Contexts

Right from the beginning, it seems useful to make a clear distinction between the two indicated contexts[5] of usage as there seems to be different reasons for relying on visual data depending on these contexts. On the one hand, visualizations are used as a kind of surrogate for the real object to ask questions about it—we will call this the "exploratory use" of visual data. On the other hand, visualizations are often used to communicate research results to scientific peers or to laymen (e.g. students, the interested public, or people working for funding institutions). Instances of the latter case we will subsequently call the "explanatory use" of visual data.[6] In both contexts, images fulfil a *mediating role*. This common feature, however, also marks the point of divergence between the two contexts as the question to be answered here is who or what the *relata* of this mediation are.

In the exploratory case, visualizations are often the results of measurement processes, i.e. the output of instruments, as in some instances direct, i.e. non-mediated, observations of the research object are not possible. The entity in question might be too small—such as molecules, atoms, electrons, etc.—or too far away—such as galaxies or quasars. Another possibility is that the feature under investigation is not accessible by human vision, for example the ultraviolet radiation of a star. Furthermore, a direct observation might be too dangerous for the scientist, for example the investigation of the interior of a damaged nuclear power plant. We may think about a variety of further possibilities, although the essential role played by visual representations basically remains the same. Taking a measurement process in its simplest form then means that the visualization relates the scientist[7] to her research object[8] and enables the latter to be investigated (safely and in accordance with the capacities of human sense perception). This implies that the scientist takes the visualization to convey certain information about the object represented. To use this image in a measurement process, the scientist must be convinced that she can *learn* something about the object from its visualization, that she can use it to test an assumption, to verify or falsify a hypothesis—in a nutshell that it can play the same *evidential role* as any other kind of datum in such a process. But what reasons support the scientist's decision to take the visualization as a surrogate here?

In the second context, the explanatory scenario, images are mediating too, although they connect different entities here than in the former case. One part of the relation is constituted by the investigating scientist wanting to communicate her research results. The other *relatum*, however, varies with the intended aim of communication. As Ludwik Fleck had already pointed out in his work about the intertwined social and epistemic processes in science, there are at least three different purposes in communicating research results (see Fleck 1986 [1936]: 86 ff.), namely *information, legitimization*, and *popularization*.[9] Whereas in the first two instances epistemic peers are related, the last one constitutes a connection between the scientist and laymen with regard to the former's profession.

Thus, on the one hand, presenting visualizations as measurement results means making the data directly accessible to one's peers and, consequently, to allow a critical assessment of one's interpretation of them. In this sense, visual data are normally presented as supporting evidence for the derived analysis. Alexander Vögtli and Beat Ernst even call this being a "visual proof" in the context of publication (see Vögtli and Ernst 2007: 72). But what reasons constitute the basis for the reliance on such data here?

On the other hand, images are often used by scientists to communicate their results to the broader public (see e.g. Kemp 2003). In the context of popularization then *cognitive as well as aesthetic considerations* play a significant role. Here the scientist has to take into consideration that her target group will not share most of her background knowledge in the relevant domain. As Fleck points out "popular science" is "science for nonexperts" (Fleck 1979 [1935]: 112). Consequently, the scientist has to *simplify* her presentation. Furthermore, she will *leave out divergent opinions*. Thus, the resulting image, text or both will *appear much more certain* than it would do in the critical discussion between peers.[10] Visualizations are, therefore, often used in a persuasive way here.[11] The scientist wants to convince the public of her results, of her opinion. And the mentioned characteristics of visual representations may better serve this purpose than a complicated written text. Consequently, scientists may even visualize data that did not originally appear in a visual form to take advantage of this persuasive mode.

Beyond such cognitive reasons, i.e. considerations to make complex research results comprehensible (or at least seemingly comprehensible) to laymen, there are often also aesthetic considerations involved in the scientist's deliberations of how to present her data. Accordingly, colors, shades, etc. are added although they were not part of the original data set (see Hennig 2009: 195–207; Groß and Duncker 2006). In this sense, Felice C. Frankel and Angela H. DePace suggest to enhance visual representations along five parameters, namely *composition, abstraction, color, layers* and *refinement* in their "practical guide to graphics for scientists and engineers" (2012).[12]

Thus, we have to take into account that recipients of such visualizations are confronted with the scientist's *deliberate decisions* on how to present her data best and, even more important, her *interpretation* of them. The choice of presentation methods may be motivated by different intentions, e.g. to enhance the comprehensibility of the data. Thus, the scientist's intentions play an essential part in this context. Consequently, images normally[13] relate laymen to the scientist's interpretation of certain data sets here, but do *not* provide access to the original research object.

In my opinion, the reservations about the epistemic status of visualizations in science expressed at the beginning of this contribution are at least partly the result of mixing up these different contexts of usage. Most of all, skeptical stances are easily derived by generalizing the case of popularization. However, this is but one aim related to the usage of visual data in science. And in the rest of this text, I want to point out that reliance on scientific images in those other contexts does not have to be an instance of epistemic credulity. On the contrary, there are a variety of reasons that we can rely on—varying, however, with the different contexts of use.

Exploratory Context: Reliance

In the exploratory context, many—though not all—scientific visualizations have their origin in *measurement processes*. Often, visual data are regarded as a kind of surrogate for the entity or process under investigation and, in this sense, become part of the scientific reasoning. Reading off the data from the image normally means being convinced that it can tell you the relevant fact about the research object, i.e. that it is a *reliable source of information*. Claiming that *Antares* is of a red color by looking at the photometric results also implies that we are not only attributing this property to the picture but also to the star itself. Yet, this does not mean that scientists literally take any property of the image to having a correlated counterpart in the object of research. Nobody, for example, would claim that *Antares* is flat just because this is a striking property of its photograph.

Wondering what might justify this epistemic practice of taking the image as a surrogate, it appears that in fact there are two correlated questions to be answered here. First, what reasons may we have to take photography as a reliable source of information? And, second, what accounts for a correct interpretation of the visual data? Thus, what allows us to read off from the picture that *Antares* is red but not flat?

Our first question is about the *instrument* and *visual data* as its output. Using photography in its detective function enables the scientist to employ a camera as a measurement device. What is crucial about the latter in general is that there exists a *causal connection* between the instrument and the object under investigation (see Harré 2010: 31), i.e. it is the object that causes the instrument to record certain data. Thus, an instrument reliably serves a detective function, *if the entity under investigation had revealed feature x, the instrument would have detected feature x (provided that it was within its range)*. In the case of photography, such a causal connection is widely acknowledged.[14] Furthermore, Maynard connects this characteristic to the context of detection. "… this is where photo-detective technologies enter the scene … Because of the remarkable sensitivity of photoreceptors, tiny chemical or electrical changes wrought on sensitive surfaces by light emissions and reflections[,] allow us to detect and to record the physical situations of their causes" (Maynard 2000: 122).

That measurement devices are reliable is usually ascertained by the *calibration* of the instrument. Calibration here means ensuring that the instrument is causally connected to the object of research in relevant respects and, thus, able to detect what it is employed for. Kelley Wilder (2009) points out that in the case of photography the main concern was about the question of how to ensure a *reliable recording* of the detected radiation, i.e. about the emulsions used. Here, scientists had to understand how the photographic picture was brought about and how it could be ensured that the required results could be reliably repeated. Hence, to obtain controlled photographic evidence, the technology of photography became an object of scientific investigation itself.[15] Once having established both a reliable photographic measurement process and the scientific understanding of it, nowadays there are methods available—based on this understanding—for the calibration of a particular camera, i.e. for ascertaining that it works in accordance with the required technical specifications.

However, the scientist's background knowledge about the causal process and its reliability will only tell her *that something has been detected* by her camera, but she will not know what. She also needs to know how her measurement results are correlated with her theory. Therefore, the second aspect that we considered relevant in epistemically judging the scientist's practice of using visualizations in their reasoning is about *interpretation*.[16] How then can we bridge the gap between theory and datum?

Laura Perini draws our attention to this very question (see Perini 2012a: 154) and suggests that what is needed here are certain "interpretative 'rules' … that are at least partly conventional" (ibid.: 163). The conventional aspect comes into play as she agrees with other philosophers that visual representations can often be used to represent different things and, therefore, also be interpreted differently (see ibid.: 166). The respective conventionality, however, is not of a totally arbitrary kind, as "[t]he selectivity of the imaging technique" (ibid.) that we can recognize as a consequence of the causal component of measurement devices will restrict possible interpretations of its visual data output.

Even though I totally agree with Perini on this constraint concerning interpretative rules, I think that we can and should be more precise about the latter—which would also mean reducing the alleged arbitrary element even further. I want to suggest that these rules should be understood as *mapping functions* that define a *kind of resemblance*[17] between visual datum and the object under investigation. Such a mapping function tells the scientist how to read off her data correctly. Moreover, this function is not an arbitrary choice in the scientific context. A great many of them are of a law-like conception. Thus, questioning the connections between visual datum and research object would also mean questioning the validity of these laws, which are normally embedded within the wider context of a theoretical network.

Take our initial example about the determination of *Antares*'s color and magnitude via photometric measurement. One important part of the mapping function in play here is that the measured brightness of the star depends on the sensitivity as a function of wavelength (bandpass) of the filter through which one observes it. A red star such as *Antares* will appear bigger in an image produced with a red filter than a hot blue star.[18] Considerations about the correlations between the age of a star and its color are then obvious components of the theoretical network in the case of photometric measurement. However, there are other hypotheses belonging to this network, too, that perhaps are not immediately apparent, such as correlations between temperatures of bodies and colors, and the like.

Beyond that, there are some mapping functions where more local definitions are at work, for example, what false colors depict in a given photograph. These decisions, however, have to be stated clearly by the interpreting scientist, e.g. in the form of captions, to ensure that others will interpret the image in the same way and thereby making the interpretation repeatable. Correspondingly, these interpretative rules are fixed for the image in question. Moreover, such local decisions can even become long-lasting conventions that are shared by the whole community.

Summing up the points so far, background knowledge about the causal connection between the instrument and the object of research, about the former's calibration, and about the resemblance between image and object, established by a relevant mapping function, constitutes reasons to rely on in the exploratory context. These reasons support the assumption that our epistemic practice of using photographic data as measurement results in scientific reasoning is indeed a rational practice. This is not to say that no errors can occur. Of course, they do. Especially in the context of basic research, there often remain problems to discern artefacts from data, as Perini pointed out (see Perini 2012c: 155). I take this, however, not to be a special problem of visualizations but regard it as a difficulty of scientific experiments and observations in general. This is also why scientists and philosophers of science take our current scientific theories not as ultimate rationales, but only as approximately true hypotheses, thus still leaving room for improvement.

Let us now turn to the second context—the explanatory context with the aim of information or legitimization—and ask whether we can discern reasons to rely on the data in this scenario, too.

Explanatory Context: Trust

In the second context, we are confronted with a *communicative act* purposefully performed by the scientist and conducted with the aid of *certain representational media,* namely by linguistic, numerical and visual representations. The scientist has the *intention* of transmitting information and wants her assertion, either written or spoken, to be taken as an *epistemic source.* Furthermore, her audience—her academic peers—normally want to *learn* something from her assertion. They want to be informed by her with regard to her research results, about her measurement methods, and her interpretation of the data. And they are convinced that they can receive this information from the scientist's communicative act.

The epistemic source put to use here by both the scientist and her audience is philosophically known as *testimony*.[19] If in such contexts visualizations are incorporated, they are normally offered as *supporting evidence* by the speaker and handled as such by her audience. Even though these visual representations can be independently assessed in the sense of questioning whether they adequately depict what they are claimed to represent by the author (see Perini 2012b: 142–7; Downes 2012: 115–30), their status as a research result—i.e. as evidence—is an effect of being part of the whole testimonial act. Looking for reasons that epistemically justify such practices, we, thus, have to consider the communicative act as a whole.

Taking into account that visualizations in the explanatory context are part of testimonial acts also means that we can make use of social epistemologists' proposals of why a recipient might be epistemically justified in accepting a given testimony. This is the place where Martin Kemp's considerations about our reasons to rely on[20] published (or otherwise transmitted) visual representations fit in (see Kemp 2012: 44). Why should we believe that there is a mammal with a beak out there just because someone said so and showed us a corresponding picture? Under what conditions is a recipient justified in accepting the scientist's testimony including the visual representation? When is she allowed to form a corresponding belief?

The debate about *knowledge by testimony* revolves around this very problem.[21] Here, I want to highlight Paul Faulkner's account of the topic (2011) that focuses on the notion of *trust*. Especially, his notion of *predictive trust* seems to be a promising one for our current purpose. "A trusts S to ø (in the predictive sense) if and only if (1) A depends on S ø-ing and (2) A expects S to ø (where A expects this in the sense that A predicts that S will ø)" (Faulkner 2011: 145). To trust the testifier in question here just means that the recipient knows about his dependency on the latter's sincerity and predicts that the speaker will act accordingly. For sure, this notion of trust alone is not sufficient to justify the acceptance of the testimony as true. "… [P]redictive trust is reasonable just when there are grounds for judging a cooperative outcome …" (ibid.). And this is exactly what we are looking for: a notion of trust that is combinable with further supporting reasons.

In the following, I will now discuss two possible sources of such supporting reasons in the scientific realm. The first one is based on the *individual* level, whereas the second is situated on the *social* level of science itself. Both aspects can be subsumed under Kemp's suggestion to take *personal experience* as a possible source of justifying reasons in the context of scientific communication (see Kemp 2003: 44).

Taking the *individual* level first, Gloria Origgi (2012) suggests the concept of *reputation* as a promising source of epistemic reasons in the context of an international and globalized scientific community. "Reputation serves the cognitive purpose of making us navigate among things and people whose value is opaque for us because we do not know enough about them" (Origgi 2012: 411). This refers to the problem mentioned in the beginning of this contribution. In a globalized community and with regard to international contributions to the scientific inquiry via journals and web-based communication, we rarely have sufficient background information about the individual researcher to assess her credibility. We plainly do not know most members of the scientific community. How can the concept of reputation be useful in this context nonetheless?

The idea is that *reputation* is a feature that spreads through a given community, i.e. although I may not know the testifier personally, others will and they will share their experience with their colleagues. The concept of reputation then allows scientific peers to learn from others about the credibility of the testifier in question. "Judgements of reputation involve always a 'third party'—a community of peers, experts or acknowledged authorities that we defer to for our evaluations. Reputation is in the eyes of the others: we look at how others look at the target and defer, with complex cognitive strategies, to this social look" (ibid.: 403).

Moreover, a scientist's reputation is also an evaluative concept, i.e. a description taking into account her former research success, but also her conformity to certain academic virtues.[22] As future success in their community may be dependent on a good reputation, it appears natural that scientists seek to achieve and to retain such a status. This also means that scientists will try to avoid behavior that would threaten their reputation, for example intentionally publishing false results. Having a good reputation, thus, will normally be an indicator of a trustworthy testifier.

What then about the second evidential source mentioned above? Here, Elizabeth Fricker (2002) made the interesting proposal that there are further empirical reasons available correlated with the *social setting of science*. "… I suggest, scientists' basis for trusting each other lies in their knowledge of each other's commitment to, and embedding within, the norms and institutions of their profession" (Fricker 2002: 383). Fricker here refers to the same social mechanism as Origgi does, namely that of creating a scientist's reputation. Contrary to Origgi, however, she does not consider it on the individual level but on the social. In this sense, the recipient in a peer-to-peer communication does not need individual knowledge about the speaker's reputation to trust her testimony; she rather adheres to the background knowledge about the mechanisms bringing about such an individual reputation in science.

Moreover, Fricker also highlights the fact that our reliance on this background knowledge is supported by further considerations about consequences of academic misconduct. Threatening one's reputation may be but one outcome of such behavior followed by more tangible ones like shortcuts of research funding or even the loss of one's job. "Unreliability is likely to be subsequently discovered and highly penalized in such a setting, and this gives one strong empirical reason, amongst others, to expect informants to be trustworthy" (Fricker 2002: 383). Here Fricker draws our attention to the fact that within the scientific community there are certain mechanisms at work to sanction such misbehavior.

Consequently, Origgi's as well as Fricker's account demonstrate sources to yield empirical reasons which allow a recipient to trust a scientist's testimony—visual representations included. This is not to deny that scientists may be mistaken. Without saying, errors can occur. However, the described social mechanisms and the threat of losing one's reputation may at least prevent deliberate cheating.

As was pointed out above, assessing the testifier's reputation and the social mechanisms creating it often belong to the *background knowledge* of the epistemic subject. Let us now turn to another possible contribution of our background knowledge to the process of credibility evaluation. This aspect is briefly mentioned by Kemp. He highlights the broadly acknowledged fact that new items learned by the word of others must be consistent with our background assumptions (see Kemp 2012: 44).[23] *Consistency* can then be regarded as a further criterion to evaluate the reliability of the testimony in question. Of course, we do not only take considerations about reputations and social roles as being part of our background assumptions but also knowledge concerning theories, instruments, calibrations, etc., i.e. all the different reasons we discussed in the exploratory context.

Consequently, there are plenty of reasons to rely on with regard to visual representations in the explanatory context. Most of them are a result of their being embedded in our epistemic practice of testimony, but some are also more specific with regard to visual representations as such.

Conclusion

As a concluding remark, let us sum up the results: We analyzed the question of what reasons may justify a scientist's epistemic practices concerning the usage of visual data in the exploratory and explanatory context. With regard to the former, we saw that the scientist's background knowledge about the instrument's causal connection to the object of research, about the former's calibration, and about the resemblance between image and object established by a relevant mapping function can constitute such reasons to be relied on. Moreover, in the explanatory context, too, there are different reasons to rely on with regard to visual data. Many of them, such as the scientist's reputation, are a result of their being embedded in our epistemic practice of testimony.

Of course, none of these reasons guarantees knowledge with certainty. Testimony is as fallible as any epistemic source. In the same way, errors may occur in the exploratory context, too. Especially in the context of basic research, there often remain problems to discern artefacts from data, as Perini pointed out (see Perini 2012c: 155). I take this, however, not to be a special problem of images but a difficulty of scientific experiments and observations in general—a fact that is reflected by scientists and philosophers of science taking our current scientific theories not as ultimate rationales, but only as approximately true hypotheses.

Acknowledgments

This work was supported by the German Research Foundation (DFG) in the context of the project *Visualisierungen in den Wissenschaften—eine wissenschaftstheoretische Untersuchung* (MO 2343/1–1). For linguistic revision I thank Janet Carter-Sigglow.

Notes

1 In the following text, I will use the expressions *visual representation*, *visualization*, *visual data*, and *image* interchangeably.
2 To see the point, note that photocopiers and production processes of computer chips belong to the same class of technologies as cameras. They all operate in accordance with the same principle (see Maynard 2000: 18, 118).
3 The next step then would be to calculate the differences between the measured magnitudes of the star in these different passbands. In the case of *Antares*, we subtract the visual magnitude (V) from its blue one (B). Additionally, we need a point of comparison. For this purpose, astronomers use the star *Vega*, whose magnitude in the UBV system has been defined as zero (see Unsöld and Baschek 2015: 180). Accordingly, *Antares* has a B-V magnitude of 1.87 in comparison to *Vega* and, according to the definition, a star is red when its color index is greater than 1.5. Consequently, *Antares* is red (see http://spiff.rit.edu/classes/phys445/lectures/colors/3, accessed December 15, 2016.
4 Kemp's example is about the discovery of the duck-billed platypus in Australia.
5 This is not meant to exclude the possibility that there may be other contexts.
6 This distinction is introduced by Frankel and DePace (2012: 13).
7 Of course, this is a simplified assumption as normally there is more than one scientist involved in the process.
8 In the following text, I will talk about the *object of investigation* in this abbreviated manner as, of course, not all topics of research are related to a concrete object but might be about processes, events or abstract entities such as species as well.
9 Another important purpose would be to use visualizations as auxiliary means in educational contexts.
10 Fleck calls this the "apodictic valuation" (Fleck 1979 [1935]: 112). He also mentions the aspect of simplification as a special feature of popular science (see ibid.).

11 Laura Perini also points out that one reason to integrate visual representations in scientific publications or talks consists in the fact that "... they play a rhetorical role and thus can help persuade an audience" (Perini 2012a: 158).

12 This implies the following: "Organize the elements and establish relationships [compose, NM]. Define and represent the essential qualities and/or meaning of the material [abstract, NM]. Choose colors to draw attention, to label, to show relationships (compare and contrast), or to indicate a visual scale of measure [color, NM]. Add layers to overlap multiple variables to create a direct relationship in physical space [layer, NM]. Edit and simplify [refine, NM]" (Frankel and DePace 2012: 15).

13 Admittedly, there are some exceptions. Most of all, there are cases where the same visualization is used in the public as well as in the scientific context. However, the assumption seems natural that this is not predominantly the case. Laymen simply lack the technical knowledge to comprehend the original data, even when presented as graphs or diagrams.

14 That this is still the prevailing way to conceive of the process of photography can be found in various contributions to Walden (2008).

15 Accordingly, Wilder speaks about the "symbiotic relationship of photography and science in the twentieth century" (Wilder 2009: 78).

16 Perini (2012a) highlights this point, too.

17 Although there are a lot of critical voices concerning the relevance of *resemblance* (see e.g. Scholz 2009: ch. 2) in the context of picture theory, I nevertheless consider this term to be appropriate in the scientific context. Last but not least, it is the conviction that datum and object share some relevant properties—which means that they resemble one another in this respect—that makes visual representations a valuable tool in scientific reasoning.

18 See Richmond, http://spiff.rit.edu/classes/phys445/lectures/colors/3, accessed December 15, 2016.

19 For a definition see Mößner (2011: 207–9) and Mößner (2010: ch. 2.4).

20 Kemp uses the term *trust* here, though I think this attitude should be reserved for human interaction. We trust people but we rely on machines, data, processes, etc. This distinction is motivated by analyses of the concept of trust in philosophy. These studies show that there is a normative dimension in the term of trust that does not appear with regard to mere reliance. Evidence to support this claim can be found in our different responses to, on the one hand, betrayed trust and, on the other hand, our coming to know that we were not justified in relying on somebody (see McLeod 2011: sec. 1).

21 For different accounts on this topic see e.g. Lackey and Sosa (2006).

22 "A reputation is the shortcut of the many strategies, heuristics and evaluations that have been placed a person or an item in a certain hierarchical configuration" (Origgi 2012: 409).

23 This point is also endorsed by Fricker, who maintains the even stronger claim that not only consistency but coherence with background beliefs is a necessary condition for justification in this case (see Fricker 2002: 380).

References

Downes, S. M. 2012. How Much Work Do Scientific Images Do? *Spontaneous Generations: A Journal for the History and Philosophy of Science* 6:*1*: 115–30.

Faulkner, P. 2011. *Knowledge on Trust*. Oxford: Oxford University Press.

Fleck, L. 1986 [1936]. The Problem of Epistemology. In: *Cognition and Fact: Materials on Ludwik Fleck*, edited by R. S. Cohen, Dordrecht [and others]: Reidel.

Fleck, L. 1979 [1935]. *Genesis and Development of a Scientific Fact.* Chicago and London: The University of Chicago Press.

Frankel, F. and DePace, A. H. 2012. *Visual Strategies—a Practical Guide to Graphics for Scientists and Engineers.* New Haven and London: Yale University Press.

Fricker, E. 2002. Trusting Others in the Sciences: A Priori or Empirical Warrant? *Studies in History and Philosophy of Science* 33: 373–83.

Groß, D. and Duncker T. H (eds.). 2006. *Farbe—Erkenntnis—Wissenschaft. Zur epistemischen Bedeutung von Farbe in der Medizin.* Berlin [and others]: Lit Verlag.

Harré, R. 2010. Equipment for an Experiment. *Spontaneous Generations: A Journal for the History and Philosophy of Science* 4:*1*: 30–8.

Hennig. J. 2009. Epistemologie des Schattens. In: *Datenbilder. Zur digitalen Bildpraxis in den Naturwissenschaften,* edited by R. Adelman, J. Frercks, M. Heßler, and J. Hennig, 195–207, Bielefeld: transcript.

Kemp, M. 2012. The Testimony of my Own Eyes—The Strange Case of the Mammal with a Beak. *Spontaneous Generations: A Journal for the History and Philosophy of Science* 6:*1*: 43–9.

Kemp, M. 2003. *Bilderwissen. Die Anschaulichkeit naturwissenschaftlicher Phänomene.* Translated by J. Blasius, Cologne: DuMont.

Lackey, J. and Sosa, E. (eds.) 2006. *The Epistemology of Testimony.* Oxford: Clarendon Press.

Maynard, P. 2000. *The Engine of Visualization—Thinking Through Photography.* Ithaca, NY: Cornell University Press.

Mößner, N. 2011. The Concept of Testimony. In: *Epistemology: Contexts, Values, Disagreement, Papers of the 34th International Wittgenstein Symposium,* edited by C. Jäger and W. Löffler, 207–9, Kirchberg am Wechsel: Austrian Ludwig Wittgenstein Society, online available at https://philpapers.org/archive/MNETCO.pdf, accessed December 21, 2016.

Mößner, N. 2010. *Wissen aus dem Zeugnis anderer—der Sonderfall medialer Berichterstattung.* Paderborn: mentis.

Origgi, G. 2012. A Social Epistemology of Reputation. *Social Epistemology* 26:*3–4*: 399–418.

Perini, L. 2012a. Image Interpretation: Bridging the Gap from Mechanically Produced Image to Representation. *International Studies in the Philosophy of Science* 26:*2*: 153–70.

Perini, L. 2012b. Truth-bearers or Truth-makers? *Spontaneous Generations: A Journal for the History and Philosophy of Science* 6:*1*: 142–7.

Perini, L. 2012c. Depiction, Detection, and the Epistemic Value of Photography. *The Journal of Aesthetics and Art Criticism* 70: 151–60.

Richmond, M. *Photometric Systems and Colors.* http://spiff.rit.edu/classes/phys445/lectures/colors/3, accessed December 15, 2016.

Scholz, O. R. 2009. *Bild, Darstellung, Zeichen—philosophische Theorien bildlicher Darstellung.* 3rd edn, Frankfurt/ Main: Klostermann.

Unsöld, A. and Baschek, B. 2015. *Der neue Kosmos. Einführung in die Astronomie und Astrophysik.* 7th edn, Berlin, Heidelberg, New York: Springer.

Vögtli, A. and Ernst, B. 2007. *Wissenschaftliche Bilder—eine kritische Betrachtung.* Basel: Schwabe.

Walden, S. (ed.) 2008. *Photography and Philosophy. Essays on the Pencil of Nature.* Malden: Blackwell Publishing Ltd.

Wilder, K. 2009. *Photography and Science.* London: Reaktion Books Ltd.

9 Pictorial Evidence: On the Rightness of Pictures

Tobias Schöttler

Pictorial Evidence, Rightness, and the Fixing of Reference

There are objects and events, both in everyday life and in scientific practices, which we know about not by direct observation but only by means of reports or images. In scientific reasoning especially, we use pictures as surrogates for the objects depicted in them, since we often do not or cannot perceive the objects themselves without technical aids such as microscopes, fMRI and so on. The images yield a specific type of evidence for the existence of the visualized entity that possesses the properties shown. In pictures we trust. We assume that they visualize real features of the depicted objects. But our sources of knowledge have to be credible. We trust other sources of knowledge (like perception or testimony) because we assume that they are able to represent the facts in question and that they usually do so. The use of scientific pictures as *pictorial evidence*, then, relies on these pictures' *rightness*.

I intentionally use the word "rightness" instead of "truth," since the correctness of pictures cannot be a matter of truth insofar as we want to hold on to a propositional understanding of truth. To apply the term "truth" to pictures would be to presuppose their propositional character. But as Nelson Goodman claims, "a picture makes no statement" (Goodman 1978: 131). In fact, a single picture makes many propositions or can be described by many different statements. Alongside the semantic ambiguity of pictures, their syntactical ambiguity is another problem for the assumption of "pictorial truth" (Goodman 1976: ch. 4; see Schöttler 2012: 254–60). In the case of natural languages, we can identify signs and differentiate between them because the alphabet helps us to distinguish between the constitutive and contingent traits of the characters. In the case of written language, the alphabet helps us to identify a unique mark as a specific letter (provided we disregard the peculiarities of the handwriting or the font). But there is no "pictorial alphabet" which we can use to distinguish between relevant and irrelevant characteristics.

The semantic and syntactical ambiguity of pictures is absent only in highly conventionalized cases such as diagrams and graphs. According to the conventions of diagrams, only specific features are relevant (for conventionalized use) while all the other features are irrelevant. Diagrams are truth-apt since they are semantically and syntactically unambiguous by convention.[1]

Other kinds of pictures, such as photographs, images gained via microscopes and many other images, do not meet these conditions of semantic and syntactic unambiguity, as Laura Perini states in her articles about visual arguments and visual confirmation (see Perini 2005a, 2005b). Her conclusion is that these kinds of pictures are not truth-apt. Since I agree, I want to use the broader concept of *rightness* to describe the correctness of pictures. In this, I follow Nelson Goodman, who uses rightness as an umbrella term, which includes truth as a special case: "Under 'rightness' I include, along with truth, standards of acceptability that sometimes supplement or even compete with truth where it applies, or replace truth for nondeclarative renderings" (Goodman 1978: 109–10).

The concept of rightness we are looking for must enable us to distinguish between representations and misrepresentations (or representations which provoke misinterpretations). More precisely, we have to ask by what means we distinguish between real properties of the "observed" object on the one hand and artefacts caused by the observation tool or the observation process on the other hand (see Hacking 1983: 200–2). In early microscopy, such artefacts were caused by aberration (see Schickore 2007: chs. 4 and 5). In the recent history of microscopy, the static view of the cell—or rather the assumption of a structured (but not a liquid) cytoplasm—can be traced back to artefacts. The assumption of a "cytoskeleton" acquired its corresponding pictorial evidence through electron microscopes.[2]

These electron microscopes showed strands between the filaments, which were interpreted as a kind of skeleton (see Wolosewik and Porter 1979; Porter and Andersson 1982). The idea of the microtrabecular lattice of the cytoplasm was made possible as a consequence of understanding these strands as real properties of the cell. Insofar as different methods of dehydration (a precondition of observation by electron microscopes) result in similar images, the idea of the microtrabecular lattice was considered to be a valid hypothesis (see Bechtel 1990). However, improvements in light microscopy revealed the structures observed to be artefacts, which are ultimately a consequence of preparation using nitrogen (or other methods of dehydration).

The thesis I am putting forward is that the distinction between real properties and mere artefacts (or in other words: between right representations and misrepresentations) is a problem for current descriptions of pictorial reference. In picture theory, two different questions are often intertwined: what is a picture and how is the reference (uniquely) fixed or determined (see Files 1996: 398–402)? Here, I am interested only in the latter question.

In this regard we can identify two main approaches in picture theory that are based on the traditional classification that distinguishes natural signs from conventional ones. The relation between natural signs and their referents is considered to be a natural one, whereas the relation between conventional signs and their referents is regarded as a non-natural one. In the history of semiotics, both the distinction and these types of signs themselves have been interpreted in various ways (see Meier-Oeser 1997).[3] The natural relation can be interpreted as a causal dependency or as a relation of resemblance; the non-natural relation can be explained by reference to some of the intentions of the users or by conventions. Augustine of Hippo, for example, explains given signs (*signa data*) by referring to the intentions of the users, and he illustrates natural signs (*signa naturalia*) by referring to causal dependencies (see Augustine 1949: 238 ff.).[4] Smoke signifies fire. Roger Bacon extends Augustine's classification by incorporating Augustine's distinctions into a broader classification (see Bacon 1978).[5] Among other things, he takes the *signa propter conformitatem* and the *signa effectus respectu causae* to be subtypes of the *signa naturalia*.

The second subtype corresponds to Augustine's understanding of natural signs. The first subtype is based on a relation of resemblance between the sign and its referent. Among his examples we find pictures. In the early modern age, semioticians and aestheticians adopted the natural/conventional distinction. They explain the meaning of conventional or arbitrary signs by reference to convention or intention. While they understand linguistic signs as conventional signs, they regard pictures as the paradigm for natural signs. With regard to natural signs, the focus in the early modern era gradually shifts from a causal relation to one of resemblance between signs and their referents.[6] Ever since Descartes's *Dioptrique* of 1637 there have been two interpretations of the relation of resemblance to consider, namely, external resemblance and internal resemblance (see Descartes 1965: 112–4). External resemblance refers to the relation between a picture and its referent, whereas internal resemblance concerns the relation between the observer's perception of the picture and of its referent.[7] (I will return to this distinction later.)

The distinction between natural and conventional signs helps us to distinguish between two kinds of theories of pictorial reference, namely, a naturalist and a conventionalist approach. Each approach implies a different concept of rightness. In this regard, it is important to note that the natural/conventional distinction implies that the reference of natural signs is determined only by the properties of the picture and what it depicts.

According to the naturalist approach, pictures are right because of their resemblance or causal relation to depicted objects. If this justification worked, we could establish an objective rightness via the relation between the picture and the depicted object (or, according to the internal resemblance, between the perception of the picture and that of what is depicted). But as I will show below, this naturalist explanation comes up against several problems.

The conventionalist approach rejects the assumption of a natural connection between pictures and their referents. As I will show, the conventionalist approach is prone to becoming a constructivist one when applied to scientific images. Thus the objects under investigation are understood as being created by their visualizations. Whereas the naturalist interpretation regards pictures as reproductions, the constructivist approach considers them as mere productions of the instruments they are observed with. Instead of objective rightness, such constructivist accounts see only a subjective justification of rightness based on the coherence between the different visualizations and the presupposed theory or symbol system. According to the natural/conventional distinction, non-natural signs can be understood in two ways. The reference of these signs can be explained either by the intentions of the users or by conventions. It is obvious that the intentionalist explanation results in mere subjectivity. For this reason, the intentionalist approaches are not suited to describe scientific images. The conventionalist approach, however, also results in subjectivity. Furthermore, such an approach is not able to address the reference objects.

Of course, the binary opposition of the naturalist and the conventionalist approach is a simplification. Picture theory encompasses a number of hybrid approaches, including perception-based accounts and intention-based accounts.[8] However, these hybrid approaches either inherit the problems of the main approaches (see below) or are not suitable when it comes to describing scientific images; the latter holds for the intention-based accounts. It is not enough for the user to intend that the picture should depict such-and-such an object or scene. Nevertheless, the binary opposition of scientific pictures as reproductions or as mere productions is not a clear-cut dichotomoy. There is a pragmatist alternative to the naturalist and constructivist approaches. Beyond the opposition of reproduction/production, we can regard pictures as vehicles that help us to individuate the depicted objects. This approach can be called a pragmatist one insofar as it takes the *use* of pictures as a starting point. Instead of the objective or the subjective justification of (a static) rightness, this model helps to establish an intersubjective justification of a dynamic rightness.

Why the Naturalist Justification of Rightness is Not Available

The naturalist approach assumes a "natural" relation between pictures and their referents. In the case of accounts based on resemblance, pictures are considered to be reproductions, whereas in the case of causal approaches, they are regarded as effects which allow us to infer the properties of the depicted object.[9] In both cases, their rightness is justified objectively by reference to the correspondence between the picture and the depicted object. This objective rightness presupposes that the pictorial reference is determined unambiguously by the resemblance or causal relation between the picture and the depicted object. If the similarity approach or the causal theory of pictures worked, we could establish an objective justification of rightness. But it is questionable whether the reference can be fixed unambiguously by a resemblance or some causal relation. After discussing the specific problems of both approaches separately (see sections "The Causal Theory of Pictorial

Reference" and "The Resemblance Theory of Pictorial Reference"), I shall sum up the consequences for the aim of establishing an objective justification of pictorial rightness (see section "A God's eye view").

The Causal Theory of Pictorial Reference

The causal theory regards pictures as traces. According to this, the relation between a picture and what it depicts is considered to be a causal relation and the depicted object is a *relevant causal factor* in the genesis of the picture (see Black 1972: 101; Scholz 2004: 82). As Max Black states: "On this view, P [the picture] might be regarded as a trace of S [the depicted] and the interpretation of P [the picture] is a matter of inference to an earlier term in a certain causal sequence" (Black 1972: 101).

The first problem is raised by the question what a "relevant factor" is. Answering this question becomes complicated because of the production process itself: In the case of the electron microscope, objects must be prepared in a special way so that they can be observed. In the case of the fMRI, the measurement data is translated into images only later, while the data is previously subjected to various processes of idealization and abstraction (see Posner and Raichle 1994: ch. 3). Even if one interprets the visual data as a trace, this data track is itself subject to several stages of transformation and post-processing. As a consequence, we have the problem of how to distinguish between the depicted object's real properties and artefacts. These artefacts are also induced by causal factors in the observation process since they are caused by the technical instrument or by the preparation of the object observed. Yet in practice we are not interested in any causal factors, only in the real properties of the objects depicted.

Pictorial evidence implies not only proof of existence but also that the picture helps us to acquire certain information about the depicted object. The second problem of this picture theory concerns the picture's content and its explication.[10] As Robert Sharpe claims: "A causal theory of representation shares with causal theories of knowledge, belief and memory one overwhelming disadvantage: it provides us with no explication of how it is that the picture has a content which 'reflects' what is represented" (Sharpe 1991: 60).

Effects (the pictures) do not necessarily tell us anything about their causes (the depicted). The causal theory of pictorial reference tells us only that the properties of the image are caused by factors entailed in the genesis of the picture. The causal theory does not even allow us to infer the visual appearance of these factors from the picture, because an effect and its cause do not necessarily resemble each other. Although a matchstick causes a fire, the fire and the match are not similar in respect of their visual appearance. At first glance, a resemblance theory promises solutions for this issue. As we shall see, however, such an approach is also confronted with several problems.

The Resemblance Theory of Pictorial Reference

We can identify two types of resemblance theories, namely, external and internal. According to external resemblance, a pictorial reference is determined by a relation of similarity between the picture and the depicted object. In order to avoid the obvious objections, some scholars focus on internal resemblance. According to this, resemblance refers to the relation between the perception of the picture and that of the depicted object. Peacocke and Hopkins, for example, refine the classic resemblance theory by combining it with a perception-based account. Although perception-based resemblance theories are able to avoid some of the problems attending the external resemblance approach, they inherit others.[11]

The use of resemblance in both approaches (external as well as internal) can be described more precisely by means of Dominic Lopes's distinction between a *representation-independent* and a *representation-dependent* resemblance. A similarity is *representation-independent* if "a similarity can be seen between a sign and its referent without first knowing its meaning" (Lopes 1996: 16). In contrast, perceiving a *representation-dependent* resemblance presupposes an awareness of the picture's reference. Since the naturalist approach necessarily presupposes that the reference is to be fixed by the resemblance, we will focus on representation-independent resemblance.

There are two main problems concerning the use of resemblance approaches to explain the objective fixing of pictorial references. First, identifying a representation-independent resemblance presupposes the ability to make a comparison between the picture and the depicted object. In many cases, this comparison is impossible when we use scientific pictures as surrogates for the depicted object, since we cannot perceive the objects themselves. This problem follows from the unobservability of the object depicted. Perception-based picture theories are confronted with a similar problem insofar as the unobservability of the object prevents us from recognizing it.[12] Second, the vagueness of the concept of similarity poses an obstacle to fixing a reference. Since every object resembles every other object in some respect, similarity alone cannot determine the reference in an unambiguous manner. Consequently, similarity is not sufficient for fixing a reference. Furthermore, similarity is not even necessary in many cases. Take, for example, the use of false colors in some scientific visualizations which depict only by convention (see Müller 2007).

It may be that the representation-dependent resemblance can explain the pictorial content in cases where the reference is fixed by other means—for example, by means of the pragmatic or linguistic context. Søren Kjørup calls this the "linguistic anchorage" of pictures (see Kjørup 1989).[13] If this approach worked, we could infer some properties of an object by analyzing a picture of it. This approach would not, in fact, solve the problem of legitimizing an objective rightness; instead, it would merely postpone the question. We would have to specify how the pictorial reference could be objectively determined by the pragmatic or linguistic context. Such a context-related fixing of reference could be explained by the intentions of the users or by conventions.[14]

Each of these strategies is jeopardized by some form of subjectivism. First and foremost, however, there would be the problem of having to distinguish between real properties and artefacts. However, the naturalist approach bases the objective justification of rightness on representation-independent and not on representation-dependent similarity. Thus we are forced to discard the objective justification of rightness, since we cannot legitimize it by relying on representation-independent resemblance. Even if the reference was fixed by a representation-independent resemblance, we would have no way of identifying such a resemblance, since we cannot compare the depiction and the depicted object without the help of the depiction.

A God's eye view?

I do not wish to deny that there are relations of causality or resemblance between some pictures and their depicted object. The question is not whether there are such correspondences; the question is whether we can determine the (specific) correspondences between the picture and the object depicted. Simply to claim the existence of correspondences is not enough. More precisely, we have to ask whether the causal theory or the resemblance approach help us to distinguish between the aspects of the picture which depict real properties and those which show artefacts.

There seems to be only one way to judge the adequacy of a depiction within the framework of these approaches, namely, by comparison. In the case of external resemblance, we have to compare the picture with the depicted object. In the case of internal resemblance, we have to compare the perceived picture with the perceived object which is depicted. In both cases, we have to know the referent beforehand.

Of course, this account necessarily presupposes a God's eye view in order to compare the depiction and the depicted, since the depicted object is often unobservable without the technical aid of optical instruments. Thus, human observers often have to rely on these technical aids. But even if the theory of correspondences by relations of causality or similarity was a correct concept of pictorial rightness, it would not provide us with a usable criterion to determine this rightness. We cannot determine if the criterion is fulfilled. Therefore, the criterion is worthless. As a consequence, objective rightness is categorically unavailable to users—and not merely temporarily beyond their grasp.

Why the Conventionalist Justification of Rightness is Not Desirable

According to the conventionalist approach, we interpret pictures relative to a unique scheme—be it a presupposed theory or a symbol system. Consequently, denotation is regarded as being "independent of resemblance"

(Goodman 1976: 5) or causal relations. Instead, for something to be a representation, a symbol system is required: "Nothing is intrinsically a representation; status as representation is relative to symbol system" (ibid.: 226). A symbol system (or, more precisely, the "scheme") correlates the symbols with their objects of reference.

Such a conventionalist approach results in constructivism. This can be exemplified by the approaches of Thomas Kuhn and Nelson Goodman (see section "Constructivist Implications"). They regard the reference objects as mere constructions of the scheme. Therefore, they can only establish a subjective justification of rightness (see section "The Subjective Justification of Rightness"), which is a very weak criterion compared to objective justification. But what is much more problematic is that a strict conventionalist approach cannot address the reference objects at all (see section "Reference Lost").

Constructivist Implications

Drawing on the work of Norwood Hanson, Thomas Kuhn posits a perceptual and a semantic theory-ladenness (see Kuhn 1970: 111–21, 127–35; Hanson 1958: ch. 1). By perceptual theory-ladenness he means that our observation (or interpretation) of visual data depends on our theories and our "conceptual web." Kuhn illustrates this claim using examples borrowed from gestalt theory (especially the gestalt switch) and by referring to the interpretation of scientific images:

> Looking at a contour map, the student sees lines on paper, the cartographer a picture of a terrain. Looking at a bubble-chamber photograph, the student sees confused and broken lines, the physicist a record of familiar sub-nuclear events.
>
> (Kuhn 1970: 111)

This implies that different observers assign different reference objects to the same picture. Accordingly, Kuhn assumes a semantic theory-ladenness: The different theories presuppose different conceptual networks, or webs, which posit differing entities (see ibid.: 102, 149). Along these lines, each theory implies some ontological commitment. Both semantic and perceptual theory-ladenness depend on a unique scheme. Such schemes can be understood as " ... ways of organizing experience; they are systems of categories that give form to the data of sensation ..." (Davidson 2001a: 183). According to this, our perception as well as our interpretation of pictures (or other symbols) are always relative to a unique scheme. "Reality itself is relative to a scheme: what counts as real in one system may not in another" (ibid.). Such a position conforms to Nelson Goodman's symbol theory, insofar as Goodman regards any ontological order and any entity as the product of a symbol system. He emphasizes:

... the object itself is not ready-made but results from a way of taking the world. The making of a picture commonly participates in making what is to be pictured. The object and its aspects depend upon organization; and labels of all sorts are tools of organization (Goodman 1976: 32).

What Goodman asserts here for pictures holds also for other means of symbolization: symbols do not refer to objects which exist independently of the symbolization. Insofar as such a conventionalist approach considers the reference objects as artefacts of the symbol systems or the symbolization itself, it can be called a constructivist one. While the naturalist approach considers pictures as reproductions, the constructivist approach regards them as mere productions.

The Subjective Justification of Rightness

If facts are fabricated and references are constructed (see Latour 1999: 15), however, what criteria remain to judge the rightness of such fabrication or construction? Constructivists decline to justify rightness by means of correspondence theory: "We cannot test a version [in other words: a representation interpreted according to a symbol system, TS] by comparing it with a world undescribed, undepicted, unperceived, but only by other means" (Goodman 1978: 4). As a consequence, constructivist approaches can only establish a subjective rightness justified by a coherence theory. Nelson Goodman proposes that representations be judged with regard to their *rightness of fit* (see ibid.: 132, 138). On the one hand, the interpretation must fit our representation practice or the symbol system; on the other hand, the representations must fit other representations of the same object.

Compared to naturalist theory's justification of an objective rightness, the subjective one is very weak. Coherence alone does not provide us with a criterion by which to distinguish right from wrong representations. Misrepresentations which provoke fallacies are not isolated exceptions. Take, for example, the numerous pictures, obtained using electron microscopy, that verify the assumption of a structured (but not a liquid) cytoplasm, and the images obtained using light microscopy, which show a liquid or at least unstructured cytoplasm (see Weiss 2012: 306 ff.). Thus, we have two intrinsically consistent sets of pictures which suggest different images of cytoplasm. Mere coherence alone cannot tell us which set represents the object adequately. As Bertrand Russell states: "[T]here is no reason to suppose that only *one* coherent body of beliefs is possible" (Russell 1914: 191). At this point, a correspondence approach would refer to the world:

> ... if there is a world independent of representation of it, as historical evidence suggests, then the aim of representation should be to describe the world, not just to relate to other representations. My argument does not

refute the coherence theory, but shows that it implausibly gives minds too large a place in constituting truth.

(Thagard 2007: 29–30)

With his reference to "minds," Thagard is suggesting that mere coherence results in a kind of idealism. This argument has at least two tacit and inter-related premises: that we have access to the objects in the world and that these objects are independent of representations. Coherentists or constructivists will likely respond by denying Thagard's premises.

The most radical constructivist reaction rejects the assumption of external referents. According to Hans-Jörg Rheinberger, there is no relation between the depiction and something outside of scientific practice (see Rheinberger 1993: 162).[15] This implies that scientific representations cannot address external referents. To disassociate the scientific practice and its representations from the world is obviously unrewarding, since we commonly want to learn something about the world by means of the sciences. Therefore, a more moderate (constructivist) response to Thagard's argument rejects his second premise that the referenced objects are independent of the representations. As illustrated, we can maintain with Kuhn and Goodman that the objects in the world depend on scientific theories or on symbol systems.[16] Of course, such a position is jeopardized by its proximity to relativism. None the less, the main problem is that the constructivists cannot address the reference of the representations used—not even the references that relate to a scheme (a theory or a symbol system) (Davidson 2001a).

Reference Lost

Every language relies on a conceptual scheme which organizes or struc-tures the world; such schemes are not translatable into one another. This is the idea of scheme relativity in a nutshell. As Donald Davidson explains in his article *On the Very Idea of a Conceptual Scheme*, the idea of scheme rela-tivity presupposes the *third dogma*, his name for the dualism consisting of an ordering scheme (like a symbol system or a scientific theory) and of unstruc-tured "neutral" content (Davidson 2001a).[17] Davidson shows that, ulti-mately, approaches which presuppose such a dualism cannot say anything about the reference and the empirical content of linguistic representations.

With some modifications, Davidson's argument can be applied to picto-rial representations: The initial question is what the conditions of possibility for understanding a picture are. In order to understand a picture in the first place, we have to identify the object as a pictorial representation. Therefore, we have to ask which criterion enables us to identify something as a picture (of something). In the case of depictions, the symbol and its content or refer-ence are inextricably linked with each other. This means that something is a depiction because it depicts something. Davidson proposes that empirical content (or the reference) can be identified by truth conditions. Applied to pictures, I prefer to speak of "conditions of rightness" (see section "Rightness as a Condition of Possibility"): What must be the case so that the pictorial representation is right or correct? One possible answer to this is that we can individuate the empirical content of the picture and can therefore justifiably consider the object as a picture. This criterion will be applicable only if we can address the world independently of the scheme.

By negating the possibility of accessing a world not structured by a symbol system or a theory, such approaches cannot explain the relation between a representation and the reality which is represented. Thus they can no longer talk meaningfully about "representations," because the concept of representation implies that something is a representation of something. They can address neither the content nor the reference of a representation. Therefore, the assumption of scheme relativity—understood as the dualism consisting of scheme and content itself—undermines the thesis of the sorting function of a scheme. Consequently, the constructivist position subverts the utility of visual data within scientific practice. Just like the radical position adopted by Rheinberger, it loses the reference of scientific pictures and reduces scientific practice to purely *formalistic play* without any link to the world. Therefore, subjective rightness is not enough.

How a Pragmatist Approach Can Resolve the Problems

The naturalist and the conventionalist approach appear to be contradictory. Each of them derives part of its plausibility from poking its finger in the problem-filled wounds of its counterpart. As long as we adhere to the dichotomy of considering pictures either as reproductions or as productions, we cannot solve the problems.

The solution is to discard the idea that we are dealing with a clear-cut contradiction here. Rather, the two alternatives simply interrelate in a contrary way. There is a third alternative which enables us to resolve the problems sketched above, namely, by subverting their presuppositions and distinctions. This approach can be labeled a pragmatist one. I understand "pragmatism" as a methodological programme which tries to clarify or resolve enigmatic or puzzling phenomena by analyzing the practices in which the phenomena in question are an issue. In contrast to the naturalist and the conventionalist approach, we can consider the data images as vehicles which help us to *individuate* their reference objects. This process of individuation can be described by Donald Davidson's model of triangulation (see section "The Triangulation of Images"). This model differs from the naturalist and the conventionalist approach not only with regard to the function of pictures, but also with regard to some of its presuppositions (see section "Rightness as a Condition of Possibility"). This pragmatist perspective entails consequences for our understanding of pictorial evidence—it gives the latter a dynamic character (see section "A Dynamic Concept of Pictorial Evidence").[18]

The Triangulation of Images

Davidson uses the metaphor of "triangulation" to describe linguistic practice as a reciprocal interpretation of agents interacting with their subject area. Reference is therefore not determined by a symbol system or by a relation of resemblance but by the people who interact with the objects in question by operating in mutual cooperation and by means of symbolization. As Davidson states:

> Without this sharing of reactions to common stimuli, thought, and speech would have no particular content—that is, no content at all. It takes two points of view to give a location to the cause of a thought, and thus to define its content. We may think of it as a form of triangulation: each of two people is reacting differentially to sensory stimuli streaming in from a certain direction. Projecting the incoming lines outward, the common cause is at their intersection. If the two people now note each other's reactions (in the case of language, verbal reactions), each can correlate these observed reactions with his or her stimuli from the world. A common cause has been determined. The triangle which gives content to thought and speech is complete. But it takes two to triangulate.
>
> (Davidson 2001b: 212–13)

Davidson's model provides us with an explanation of reference fixation. In the course of the triangulation process, the individuals involved in communication generate preliminary symbol systems or tentative theories. Unlike the constructivist or conventionalist approach, a symbol system or theory understood in this way is something that is modified in each case and each practice through the reciprocal adjustment of actions by the participants in the communication. The individuals involved adjust their particular symbol systems in each interaction in order to communicate with each other. This communication is enabled by the common stimuli.

Davidson's model of triangulation is obviously tailored to linguistic communication. But because of its very general character, it can also be used to describe how we deal with pictures (see Vogel 2003: 107–34, 213–5, especially 130–3). When applying the triangulation model to images, the basic concepts of the approaches discussed above are adopted but take on completely different meanings. As already mentioned, the observers employ theories or symbol systems. But unlike the conventionalist approach, the pragmatist approach considers them as provisional tools which have a dynamic character insofar as they are constantly being modified.[19] Most notably, triangulation helps us to *individuate* the reference object, but it does not *create* the reference object as assumed by the constructivist approach. In the same way, the pragmatist approach can make use of the concept of similarity. I use the word "similarity" here as an umbrella term for all types of correspondences which might be relevant. However, similarity here acquires a different function. Although we cannot expect a representation-independent resemblance to determine the reference, we can use a representation-dependent resemblance to gain information about the depicted object. Like the symbol systems, the adoption of such similarity is only a provisional tool.

Since the pragmatist approach regards pictures as a means to individuate their reference objects, it subverts the distinction between reproduction and mere production. However, the main difference between the pragmatist approach and both the naturalist and conventionalist ones pertains to the character and function of rightness.

Rightness as a Condition of Possibility

For Davidson, the condition of possibility for triangulation is the *principle of charity*. In the case of linguistic communication, this means that we have to presuppose that the utterance of the other participant in the communication is true. Otherwise, we have no chance to understand him or her. Davidson uses the word "truth" in a very broad sense; it implies criteria of relevance, of internal consistency, of rationality and so on. As mentioned at the start of this piece, I want to use the concept of rightness rather than the concept of truth to describe pictures. Following Nelson Goodman, we can understand rightness as *rightness of fit*:

> Such matters aside, a statement is true, and a description or representation is right, for a world it fits. And a fictional version, verbal or pictorial, may if metaphorically construed fit and be right for a world. Rather than attempting to subsume descriptive and representational rightness under truth, we shall do better, I think, to subsume truth along with these under the general notion of *rightness of fit*.[20]
>
> (Goodman 1978: 132, emphasis added)

We certainly check a picture's rightness by examining if the picture is adequate to our purposes and fits other visualizations of the same object. Goodman focuses on this use of rightness for a subsequent validation of symbolizations. But to come into play, i.e. to understand a representation, we have to consider rightness as a precondition for the understanding of scientific pictures too. In order to understand the pictures in use, we have to assume that they are adequate representations of something. Without presupposing this, we are not justified in considering the object as a picture and we are not able to determine its empirical content.

Using rightness as an antecedent presupposition, we invert the directionality of the relation between representation and rightness. This leads to consequences for the "skeptic challenge." Skeptical positions maintain a general suspicion regarding potential deception, often rooted in the possibility of manipulating pictures by means of Photoshop and other computer software (see Mitchell 1994: 30–1; Savedoff 2008: 134 ff.). The focus on

Photoshop and digital pictures certainly seems to be overstated. The majority of pictures where deceptions have come to light were not caused by such manipulations.[21] What is much more important is that the assumption of deception does not convincingly follow from its mere possibility. The main problem inherent in such blanket skepticism, however, is that it allows no room for us as users to enter the equation. Starting from a global skepticism, we cannot come into the play (of understanding pictures). As Wittgenstein states, "If you tried to doubt everything you would not get as far as doubting anything. The game of doubting itself presupposes certainty" (Wittgenstein 1972: § 115).[22]

In this sense, the Cartesian question is subverted. The skeptical challenge is whether we have a reason not to doubt. In contrast to this, the pragmatist question is whether we have a reason to doubt. Insofar as the triangulation model presupposes the rightness of pictures, it subverts such global skepticism. Even so, it does not result in a naïve belief in a given representation; rather, it allows for local skepticism. This can be exemplified by Fox Mulder and Dana Scully—the main characters in the long running TV series *The X Files*. In the early episodes, they represent opposing attitudes. Mulder is the believer and Scully is the skeptic. Scully often cannot make sense of photographic evidence; that is, she cannot interpret it. Mulder can interpret it by provisionally presupposing its rightness. Although Mulder wants to believe (fans will remember the poster pinned up in his office), he is not a naïve believer. Equipped with local skepticism, he discloses some faked evidence.

A Dynamic Concept of Pictorial Evidence

The triangulation model subverts the possibility of *global skepticism* insofar as assumed rightness is the condition of possibility for the triangulation or for understanding a picture. Although the triangulation model presupposes the rightness of a picture, it equally enables *local skepticism*: "[I]t seems clear that a belief of any kind, true or false, relies for its identification on a background of true beliefs; for a concept, like that of a mouse or chair, cannot remain the same concept no matter what beliefs it features in" (Davidson 2001b: 195).

We can identify disturbances and misinterpretations by detecting inconsistencies within the pictorial representation or between different representations. But this presupposes the (provisional) assumption of their possible rightness. Comparing different visualizations enables us to identify which characteristics are caused by the object itself and which are caused by the technical preparation of the object or by the optical instrument. This takes the presence of technical aids and methods into account. We do not identify the pictures of the microtrabecular lattice as artefacts because we know the real objects. Rather, we identify them as artefacts because we know that dehydration is a precondition for using electron microscopes.

In the triangulation model, the question of (pictorial) rightness is postponed. We do not seek any guarantee that the pictures are correct representations. Instead, we presuppose that they are right, and then we ask if there are reasons to doubt this in order to identify misrepresentations.

One might object that the conventionalist account could also introduce the convention that the pictures represent the depicted objects correctly. However, this strategy would not work, as the static concept of convention presupposed by strict conventionalism does not allow for any modification (see Schöttler 2012: 270, 278–9). Instead, we must substitute one convention with another; in doing so, we inherit the problems of Kuhn's paradigm shift: Because of the relativism and the incommensurability implied by his concept of paradigm, Kuhn cannot really explain the paradigm shift. But the question is: why should we do this in the first place? Why should we want to change the scheme? If the objects of reference are relative to a scheme, there cannot be any conflict between the scheme and the world since the conventionalist approach still has the problem of addressing the objects of reference independently of the scheme (see section "Reference Lost").

In contrast to this, triangulation takes place between the observers involved in interaction with the subject area, in order to individuate the objects. All conventions are only presupposed in a tentative manner and can therefore be modified during their application. Since rightness is presupposed, the question cannot be "which representations are right?" Instead, the question must be "which representation cannot be right?" and "how do we have to modify our presupposed assumption of rightness accordingly?" Consequently, we have to understand rightness as dynamic and tentative. From this, it follows that we have to understand pictorial evidence as dynamic and tentative too. Insofar as objective justification is not available and mere subjective justification is not desirable, we are left with only tentative rightness and tentative pictorial evidence.

Notes

1 We can distinguish diagrams from other kinds of pictures by means of Nelson Goodman's concept of *relative repleteness* (see Goodman 1976: 230). Unlike other kinds of pictures, diagrams are *relatively attenuate* since there are fewer aspects constitutive or relevant for their use than in the case of photographs or paintings. Therefore, the diagram's semantic and syntactical ambiguity is restricted.

2 For studies that relate the cytoskeleton theory to the history of microscopy see Breidenmoser et al. (2010: 20–1, 59–60) and Weiss (2012: 306 ff.).

3 The distinction and the *principia divisionis* used to establish it are essentially contested (see Rollin 1976).

4 For a more detailed analysis see Meier-Oeser (1997: 24–30). Meier-Oeser offers cogent reasons for questioning the translation of *signa data* as conventional signs.

5 In the absence of space to analyze Bacon's classification in detail, the reader is referred to Howell (1987) and Meier-Oeser (1997: 54–9).

6 Concerning the following see Dubos (1967: 365 ff.); Harris (1783: 72–1) and Lessing (1987: 608–11). Concerning Dubos, Harris and Lessing see also Schöttler (2012: ch. III.2).

7 More recent proponents of an internal resemblance theory are Christopher Peacocke and Robert Hopkins (see Peacocke 1987; Hopkins 1995; see also Steinbrenner 2009: 290).

8 Peacocke and Hopkins vindicate a perception-based resemblance theory whereas Lopes combines a perception-based approach with a conventionalist one (see Lopes 1996). Novitz and Kjørup defend intention-based accounts (see Novitz 1977; Kjørup 1978). Novitz bases his resemblance theory on an intention-based approach, whereas Kjørup combines an intention-based account with a symbol theory inspired by Goodman.

9 With regard to causal interpretation see Bunge (2010).

10 On further problems associated with the causal theory of pictures see Black (1972: 101–3) and Scholz (2004: ch. 3).

11 For the problems associated with the resemblance theory in general see Scholz (2004: ch. 2) and Schöttler (2012: 168–92). For a more general critique of the internal resemblance theories of Peacocke and Hopkins see McIntosh (2003) and Schöttler (2012: 177–84).

12 See for example Peacocke (1987) and Lopes (1996).

13 The use of visual data bears a likeness to linguistic testimony. In many cases of testimony, reductionism fails insofar as we cannot check the testimonial knowledge by using other sources of knowledge (see Lackey 2006a; 2006b). In many cases of visual data we cannot check the depiction by comparing it with the depicted object itself—unless we are equipped with a God's eye view.

14 For intention-based accounts see Kjørup (1987) and Novitz (1977). For a critical discussion of convention-based accounts, see "How a Pragmatist Approach Can Resolve the Problems".

15 Rheinberger is obviously inspired by Jacques Derrida's semantics. According to the latter, nothing exists outside the text: "Il n'y a pas de hors-texte" (Derrida 1967). For Derrida's semantics see Derrida (1967: 409–28).

16 For example Whorf (1967: 207–19); Quine (1951: 20–43); Kuhn (1970: 102–3, 145–50). Feyerabend (1975) and Goodman (1978) hold different versions of this assumption. Mößner offers a critique of the constructivist approach to scientific pictures, from a realistic point of view (see Mößner 2013).

17 See also Bertram, Lauer, Liptow, and Seel (2008: 151 ff.).

18 István Danka regards the dynamic relation between reality and knowledge as an important characteristic of pragmatist approaches (see Danka 2013: 34).

19 Conventionalist approaches usually imply a static character of symbol systems (see Schöttler 2012: 268 ff., 275 ff.).

20 See also ibid., 138.

21 See the examples of pictures that deceive in (Haus der Geschichte der Bundesrepublik Deutschland 2003).

22 See also Pritchard *forthcoming*.

References

Augustine. 1949. De doctrina Christiana. In: *Oeuvres de Saint Augustin, vol. 11:* translated and introduced by G. Combés and J. Farges, 168–541, Paris: Desclée De Brouwer.

Bacon, R. 1978. An Unedited Part of Roger Bacon's Opus Maius: De Signis. edited by K. M. Fredborg, L. Nielsen and J. Finborg, *Traditio* 34: 75–136.

Bechtel, W. 1990. Scientific Evidence: Creating and Evaluating Experimental Instruments and Research Techniques. *PSA: Proceedings of the Biennial Meeting of the Philosophy of Science Association* 1: 559–72.

Bertram, G. W., Lauer, D., Liptow, J., and Seel, M. 2008. *In der Welt der Sprache: Konsequenzen des semantischen Holismus.* Frankfurt/Main: Suhrkamp.

Black, M. 1972. How Do Pictures Represent? In: E. H. Gombrich, J. Hochberg and M. Black. *Art, Perception, and Reality*, 95–130, Baltimore/London: John Hopkins University Press.

Breidenmoser, T., Engler, F. O., Jirikowski, G., Pohl, M., and Weiss, D. G. 2010. *Transformation of Scientific Knowledge in Biology: Changes in our Understanding of the Living Cell through Microscopic Imaging.* Berlin: Max-Planck-Institut für Wissenschaftsgeschichte.

Bunge, M. 2010. Reading Measuring Instruments. *Spontaneous Generations: A Journal for the History and Philosophy of Science* 4: 85–93.

Danka, I. 2013. How to Do Things with Perception: Enactivism and Pictorial Representation. In: *Enacting Images: Representation Revisited*, edited by Z. Kondor, 13–46, Cologne: Halem.

Davidson, D. 2001a [1974]. On the Very Idea of a Conceptual Scheme. In: D. Davidson, *Inquiries into Truth and Interpretation*, 183–98, Oxford/New York: Oxford University Press.

Davidson, D. 2001b [1991]. Three Varieties of Knowledge. In: D. Davidson, *Subjective, Intersubjective, Objective*, 205–220, Oxford/New York: Oxford University Press.

Derrida, J. 1967. La structure, le signe et le jeu dans le discours des sciences humaines. In: J. Derrida, *L'écriture et la différence*, 409–28, Paris: Édition du Seuil.

Derrida, J. 1967. *De la Grammatologie.* Paris: Édition du Seuil.

Descartes, R. 1965. La Dioptrique. In: R. Descartes, *Œuvres de Descartes, vol. VI: Discours de la méthode & essais*, edited by C. Adam and P. Tannery, 79–228, Paris: Vrin.

Dubos, J.-B. 1967. *Réflexions critique sur la poésie et sur la peinture.* Vol. I, Genève: Slatkine.

Feyerabend, P. 1975. *Against Method: Outline of an Anarchistic Theory of Knowledge.* London: NLB.

Files, C. 1996. Goodman's Rejection of Resemblance. *British Journal of Aesthetics* 36: 398–402.

Goodman N. 1978. *Ways of Worldmaking.* Indianapolis, Indiana: Hackett Publishing Company.

Goodman N. 1976. *Languages of Art. An Approach to a Theory of Symbols.* 2nd edn, Indianapolis/Cambridge: Hackett.

Hacking, I. 1983. *Representing and Intervening: Introductory Topics in the Philosophy of Natural Science.* Cambridge: Cambridge University Press.

Hanson, N. 1958. *Patterns of Discovery.* Cambridge: Cambridge University Press.

Harris, J. 1783. *Three Treatises: The First Concerning Art. The Second Concerning Music, Painting, and Poetry. The Third Concerning Happiness.* [first edition in 1744] London.

Haus der Geschichte der Bundesrepublik Deutschland (ed.) 2003. *Bilder, die lügen.* 3rd edn, Bonn: Bouvier.

Hopkins, R. 1995. Explaining Depiction. *The Philosophical Review* 104:3: 425–55.

Howell, K. 1987. Two Aspects of Roger Bacon's Semiotic Theory in De Signis. *Semiotica* 63: 73–82.

Kjørup, S. 1989. Die sprachliche Verankerung des Bildes. *Zeitschrift für Semiotik* 11:*4*: 305–17.

Kjørup, S. 1978. Pictorial Speech Acts. *Erkenntnis* 12: 55–71.

Kuhn, T. S. 1970. *The Structure of Scientific Revolutions.* 2nd edn, Chicago: University of Chicago Press.

Lackey, J. 2006a. Introduction. In: *The Epistemology of Testimony*, edited by J. Lackey and E. Sosa, 1–21, Oxford: Oxford University Press.

Lackey, J. 2006b. It Takes Two to Tango: Beyond Reductionism and Non-Reductionism in the Epistemology of Testimony. In: *The Epistemology of Testimony*, edited by J. Lackey and E. Sosa, 160–89, Oxford: Oxford University Press.

Latour, B. 1999. *Pandora's Hope: Essays on the Reality of Science Studies.* Cambridge, Mass.: Harvard University Press.

Lessing, G. E. 1987. An Friedrich Nicolai. Hamburg, d. 26. Mai 1769. In: *Werke und Briefe. Bd. 11/1. Briefe von und an Lessing 1743–1770*, edited by H. Kiesel, 608–11, Frankfurt/ Main: Deutscher Klassiker Verlag.

Lopes, D. 1996. *Understanding Pictures.* New York: Oxford University Press.

McIntosh, G. 2003. Depiction Unexplained: Peacocke and Hopkins on Pictorial Representation. *British Journal of Aesthetics* 43:*3*: 279–88.

Meier-Oeser, S. 1997. *Die Spur des Zeichens. Das Zeichen und seine Funktion in der Philosophie des Mittelalters und der frühen Neuzeit.* Berlin: de Gruyter.

Mitchell, W. J. 1994. *The Reconfigured Eye: Visual Truth in Post-Photographic Era.* Cambridge/London: MIT Press.

Mößner, N. 2013. Photographic Evidence and the Problem of Theory-Ladenness. *Journal for General Philosophy of Science* 44:*1*: 111–25.

Müller, S. 2007. Visualisierung in der astronomischen Digitalfotografie mit Hilfe von Falschfarben. In: *Vom Bild zur Erkenntnis? Visualisierungskonzepte in den Wissenschaften*, edited by D. Groß and S. Westermann, 93–110, Kassel: Kassel University Press.

Novitz, D. 1977. *Pictures and their Use in Communication: A Philosophical Essay.* Den Haag: Nijhoff.

Peacocke, C. 1987. Depiction. *The Philosophical Review* 96:*3*: 383–410.

Perini, L. 2005a. The Truth in Pictures. *Philosophy of Science* 72: 262–85.

Perini, L. 2005b. Visual Representation and Confirmation. *Philosophy of Science* 72: 913–26.

Porter, K. and Andersson, K. 1982. The Structure of the Cytoplasmic Matrix Preserved by Freeze-drying and Freeze-substitution. *European Journal of Microscopy* 29: 319–32.

Posner, M. I. and Raichle, M. E. 1994. *Images of Mind.* New York: Sci Am Library.

Pritchard, D. *forthcoming.* Wittgenstein on Hinges and Radical Scepticism. In: *On Certainty: Blackwell Companion to Wittgenstein*, edited by H.-J. Glock and J. Hyman. Hoboken: Blackwell.

Quine, W. V. O. 1951. Two Dogmas of Empiricism. *The Philosophical Review* 60:*1*: 20–43.

Rheinberger, H.-J. 1993. Vom Mikrosom zum Ribosom. "Strategien" der "Repräsentation" 1935–1995. In: *Die Experimentalisierung des Lebens. Experimentalsysteme in den biologischen Wissenschaften 1850/1950*, edited by H.-J. Rheinberger and M. Hagner, 162–87, Berlin: Akademie-Verlag.

Rollin, B. E. 1976. *Natural and Conventional Meaning: An Examination of the Distinction.* Den Haag: Mouton.

Russell, B. 1914. *Problems of Philosophy*. London: Williams and Norgate.

Savedoff, B. 2008. Documentary Authority and the Art of Photography. In: *Photography and Philosophy: Essays on the Pencil of Nature*, edited by S. Walden, 111–37, Malden, MA: Blackwell.

Schickore, J. 2007. *The Microscope and the Eye: A History of Reflections, 1740–1870*. Chicago/London: University of Chicago Press.

Scholz, O. R. 2004. *Bild, Darstellung, Zeichen. Philosophische Theorien bildlicher Darstellung*. 2nd edn, Frankfurt/Main: Klostermann.

Schöttler, T. 2012. *Von der Darstellungsmetaphysik zur Darstellungspragmatik. Eine historisch-systematische Untersuchung von Platon bis Davidson*. Münster: Mentis.

Sharpe, R. A. 1991. *Contemporary Aesthetics*. Aldershot: Gregg Revivals.

Steinbrenner, J. 2009. Bildtheorien der analytischen Tradition. In: *Bildtheorien: Anthropologische und kulturelle Grundlagen des Visualistic Turn*, edited by K. Sachs-Hombach, 284–315, Frankfurt/Main: Suhrkamp.

Thagard, P. 2007. Coherence, Truth and the Development of Scientific Knowledge. *Philosophy of Science* 74: 28–47.

Vogel, M. 2003. Medien als Voraussetzungen für Gedanken. In: *Medienphilosophie: Beiträge zur Klärung eines Begriffs*, edited by S. Münker, A. Roesler, and M. Sandbothe, 107–34 and 213–5, Frankfurt/Main: Fischer.

Weiss, D. G. 2012. Das neue Bild der Zelle: Wechsel der Sichtweisen in der Zellbiologie durch neue Mikroskopieverfahren. In: *Visualisierung und Erkenntnis. Bildverstehen und Bildverwenden in Natur- und Geisteswissenschaften*, edited by D. Liebsch and N. Mößner, 295–328, Cologne: Halem.

Whorf, B. L. 1967 [1940]. Science and Linguistics. In: *Language, Thought, and Reality: Selected Writings of Benjamin Lee Whorf*, edited by J. B. Carroll, 3rd edn, 207–19, Cambridge, Mass.: M.I.T. Press.

Wittgenstein, L. 1972. *On Certainty*. New York: Harper & Row.

Wolosewik, J. J. and Porter, K. R. 1979. Microtrabecular Lattice of the Cytoplasmic Ground Substance: Artifact or Reality. *Journal of Cell Biology* 82: 114–39.

Part III

Measuring the Immeasurable

10 Measurement in Medicine and Beyond: Quality of Life, Blood Pressure, and Time

Leah McClimans

Quality of life measures are popular with health policymakers in large part because of their ability to function as quantitative measuring instruments, while also providing the patients' point of view (see Darzi 2008). From a development perspective this attraction requires that these measures are epistemically and ethically sound. This double burden has proven difficult to achieve and these instruments have received significant criticism, mostly from those who develop and work with them. For instance, in 1995 the *Lancet* ran an editorial cautioning the use of these measures as end points in clinical trials and in 1997 Sonia Hunt's editorial in *Quality of Life Research* argued that they are misleading and probably unethical; more recently in 2007 Jeremy Hobart and colleagues argued in *Lancet Neurology* that almost all current measures are invalid (see ibid.). In my own work I have argued that they are invalid and difficult to interpret, at least in part because they do not accurately represent the patients' point of view (see McClimans 2010a; 2010b; 2011).

In this contribution I ask why quality of life measures face these challenges. One explanation that researchers commonly invoke is that quality of life measures lack a "gold standard" and are thus more difficult to measure than physical properties such as blood pressure. In what follows I examine and reject this explanation and offer a different one: the problems that quality of life measures encounter arise because quality of life lacks a theory that provides a representation of the measurement interaction, i.e. the relationship between the quality of life construct and its instruments.

Gold Standards

Hunt begins her editorial in *Quality of Life Research* by arguing that the main problem with measuring quality of life is that there is no "gold standard" (1997: 206). Quality of life researchers use this phrase to refer to the fact that these instruments lack external criteria against which outcomes of quality of life could be compared and tested. Another way to put the point is to say that the quality of life construct is not observable (see Streiner and Norman 2008: 178–9; Bridgman 1980).

The absence of a gold standard is often used to explain certain features of quality of life measurement. For instance, Streiner and Norman in their book *Health Measurement Scales* link the need to construct-validate quality of life instruments with their lack of observable content (2008: 178). Hobart and colleagues allude to the lack of a gold standard to explain why quality of life must be measured indirectly (see Hobart et al. 2007: 1095). For her part, Hunt connects this absence to (1) problems interpreting quality of life outcomes, i.e. determining the clinical significance of these outcomes, and (2) the plethora of different questionnaires that are used to measure quality of life (see Hunt 1997: 206).

In terms of her first concern, Hunt argues that the lack of a gold standard prohibits our ability to determine a normal value range for quality of life, i.e. the parameters within which quality of life is acceptable. Hunt seems to believe that external criteria can typically help to obtain an agreement about the meaning of a construct and thus what constitutes a normal or average value range. But the absence of a gold standard makes agreement on a normal range of quality of life difficult to obtain. Moreover, without such a consensus Hunt argues that the construction of quality of life measures and the interpretation of their outcomes proceed in an ad hoc manner.

This manner of instrument construction leads to her second concern that there is a surfeit of poorly conceptualized quality of life measures. She suggests, moreover, that the lack of a gold standard has given researchers an unconscious license to measure quality of life through whatever means they fancy or find convenient (see Hunt 1997: 206). Her claim seems to be that without external criteria to guide conceptual development and measurement design questionnaires proliferate without restraint. This practice has in turn led to the development and use of invalid and unethical instruments (see ibid.: 205).

Our inability to determine a normal value range for quality of life—a range such as we have for blood pressure and BMI—and the multitude of unrelated measures that are called "quality of life" suggest to Hunt that we do not know what this construct *is*. Repeatedly, Hunt states that if we do not know what something is, we cannot measure it and we should not use it to make decisions about treatment. Moreover, she argues if we want to understand quality of life better, then we would do well to turn to philosophy. For, as Hunt writes, quality of life belongs "... perhaps most clearly to the realm of moral philosophy" (ibid.: 210).

For most quality of life researchers, the lack of a gold standard points to certain problems to which the measurement of quality of life must overcome or succumb. For many, overcoming this lack is a matter of determining the appropriate scale development and analysis. Put differently, the measurability of the construct is not questioned, it is presupposed. Hunt argues, however, that the lack of a gold standard should give us pause to consider the nature of quality of life. *Why* does quality of life lack a gold standard? For Hunt part of the answer is that quality of life is deeply bound up with idiosyncratic values and dynamic interactions across a number of dimensions, e.g. physical functioning, emotional health, social life (see ibid.: 206). It simply does not make sense to think of quality of life as having a standard.

A Psychometric Response

In her 2012 conference poster at the International Society for Quality of Life's annual meeting, Carolyn Gotay responds to suggestions such as Hunt's that quality of life is a philosophical construct and, as she interprets the consequences of this claim, immeasurable.[1] Gotay argues that quality of life cannot be immeasurable because (a) we have face and content valid instruments and (b) to the extent that quality of life measures encounter challenges these challenges are no different than the kind of challenges facing more "concrete" measures such as blood pressure. I am most interested in Gotay's second response, but before moving on to discuss it I want to examine briefly her first one.

Validity

Gotay claims that we currently have valid quality of life measures. She uses this fact to deny that quality of life cannot be measured. Although this position is common to quality of life researchers, it is overly simplistic. Do we have positive validation studies for many quality of life measures? Yes. Are these validation studies themselves valid? In many cases, the answer is no (see McClimans 2010a: 225–40; Hobart et al. 2007: 1100).

An increasingly common criticism of construct validation in the context of classical test theory is its inability to determine if a measure represents its object of inquiry, i.e. its inability to provide evidence of validation (see Hobart et al. 2007: 1100–1). There are two kinds of construct validity, internal and external (see Streiner and Norman 2008: 174–5). External validity provides information about how a new instrument relates to established instrument, but it does not tell us what the instrument under consideration measures. Internal construct validity has a similar problem: it tells us how well the items on a scale are related to one another, but it does not tell us what these items represent. To be sure, Gotay also mentions face validity. This source of validation asks experts to give their opinion regarding how well the items on a scale represent the target construct. Although face validity is properly aimed at assessing the relationship between the instrument and its object of inquiry, most researchers do not believe that expert opinion is sufficient to establish validity. Face validity is taken to be subjective and open to bias (see Salzberger 2013: 2–3).

Contrary to Gotay's response, the fact that we have hundreds of validated quality of life instruments does not speak in favour of quality of life's measurability. Indeed, some have argued that the extent of positive validation studies ranging over so many different quality of life measures undermines their claim to validity (see ibid.: 2). In any case the lack of a valid validation method does not indicate that we cannot measure quality of life, only that validation studies using this method cannot provide evidence in favour of measurability. But what of Gotay's claim that quality of life instruments are similar to other more "concrete" instruments such as blood pressure?

Confounding Variables

In the article that Gotay cites on blood pressure the authors are concerned with the accuracy of blood pressure readings particularly those that are taken at the clinic. Clinically obtained blood pressure readings are confounded by short-term biological fluctuations such as those that occur from white coat effect, e.g. anxiety. Providers have identified uncertainty about patients' true blood pressure as one of the most common reasons for not treating patients or not changing therapy when they present in the clinic with high blood pressure (see Kerr et al. 2008).

In this study researchers compared numerous blood pressure readings in three different settings (research, home and clinic) over the course of 18 months. They found that a significant cause of the inaccuracy of single readings comes from unavoidable within-patient variation of blood pressure over time such as short-term anxiety. The authors conclude that individualized algorithms that account for each patient's variability and/or calculated averages from home blood pressure measures are needed to improve the accuracy of blood pressure outcomes and clinical decision-making (see Powers et al. 2011: 785–6).

Confounding variables such as anxiety are common in measurement, and as we will see when we examine the measurement of time so too are the algorithms and models used to account for these variables (see van Fraassen 2010: 166). Gotay's claim is that to the extent that quality of life faces measurement challenges they are challenges that almost all measures face. We should not think that quality of life is immeasurable simply because variables threaten to confound our outcomes.

Within quality of life research one important source of bias or confounding that has received significant attention in the last decade is the phenomenon of response shift. Response shift is defined as a change in the meaning of one's self-evaluation of a target construct as a result of (a) a change in the respondent's internal standards of measurement, (b) a change in the respondent's values and (c) a redefinition of the target construct (see Schwartz and Sprangers 1999: 1532). Response shift is important because it begins to articulate more precisely at least part of Hunt's conclusion that quality of life is dynamic, idiosyncratic and value-based.

Here are some examples of response shift: (1) respondents change their understanding of what constitutes a good quality of life from one of physical activity to one consisting of time with family and friends; (2) over time a respondent alters her understanding of what counts as a 6 on the 1–10 pain scale. To use Hunt's language these changes are dynamic, may be idiosyncratic, and often involve values. But the difference between Hunt and many other quality of life researchers is that while Hunt takes this dynamism and value-base as inherent to the quality of life construct (and thus part of what makes it philosophic) other researchers such as Gotay, for example, conceptualize response shift as bias (we need to find a way to measure and correct for it).

Response Shift

In their paper *Formal definitions of measurement bias and explanation bias clarify measurement and conceptual perspectives on response shift* Oort et al. describe the threat that response shift poses when it is conceptualized as bias: "Observed changes in respondents' test scores may reflect something other than true changes in the attribute that we want to measure" (Oort et al. 2009: 1126). To better understand this threat, consider that measurement bias is defined as a violation of measurement invariance. Measurement invariance says that if we have a latent construct such as quality of life that is measured by the scales on a questionnaire, then the instrument is unbiased with respect to a set of variables, e.g. age, sex, disease, etc. if the principle of conditional dependence holds (see ibid.: 1128). In other words, given a score of quality of life on our questionnaire, none of these other variables, i.e. age, sex, disease, etc. should help to explain that score, i.e. only quality of life should explain the score.

Response shift, however, violates measurement invariance because given repeated administrations of a quality of life instrument variables such as one's values, personal identity and experience with a disease or health state *do* affect the explanation of the score (see ibid.: 1130). To put this point differently, however, researchers define quality of life (and different instruments operationalize the construct differently) the description of response shift as measurement bias presupposes that this definition does not include values, identity or personal experience with one's disease or health state.

To be clear about the consequences of conceptualizing response shift as measurement bias, consider an example. You grow up in Western Pennsylvania with four brothers, weekends hunting with your siblings and training for marathons is your version of a good life. Then somewhat suddenly you are beset with chronic allergies and joint replacements. Marathons and hunting become difficult and for some time you are depressed and the quality of your life is poor. Then your daughter has a child and spending time with your granddaughter and other members of your family begins to fill the gap that your inability to hunt and run had left. You begin to see the value in a different kind of life, a more social life, a life that attends to the details of the lives of those you love. As a result, your quality of life improves and if asked to compare your life now with the quality of life you had before your illnesses you would say that this one was better because it was richer.

From a clinical perspective this example represents a success, i.e. we want patients to adapt to their circumstances. From a measurement perspective this adaptation is a response shift and thus obscures true changes in the attribute we want to measure, in this case quality of life. Many have found this assumption offensive and prejudicial. To say that my reported quality of life is biased because I have come to value things that I did not value before has struck some—especially those within the disability community—as ableist (see Wasserman et al. 2005: 11–125; McClimans et al. 2013: 1875–6; McClimans 2010b: 70–6).[2]

In response to this criticism, those who conceptualize response shift as measurement bias are careful to qualify descriptions such as "bias" and "confounding" as purely statistical. They emphasize that although these terms are usually understood as pejorative, response shifts themselves can be advantageous to individuals and groups who experience them (see Oort et al. 2009: 1135). Thus researchers use the measurement perspective, but emphasize the clinical one. The coherency of this response depends to a large degree on the justification for treating response shift as a confounding variable, thus I want to focus on evaluating our conceptualization of measurement accuracy in blood pressure and quality of life.

It is with regard to the challenges that confounding variables present that Gotay suggests quality of life measures are similar to measures of blood pressure. Yet while most find it convincing that increased anxiety at the doctor's office and other experiences throughout the day can obscure one's true blood pressure reading; it is less convincing that response shifts obscure the true value of quality of life. Something being less convincing does not make it untrue, but to be able to evaluate this comparison we need to understand better the relationship between accuracy, measuring instruments and the variables that can confound them. In the next section I turn to some of the recent literature in the epistemology of measurement in the physical sciences to explore these relationships.

Measuring Time

In his article *How Accurate Is the Standard Second*, Eran Tal presents an account of measurement accuracy in the context of time. Specifically, he addresses the accuracy of the 13 atomic clocks or "primary standards" around the world that are used to "realize" the definition of the second (see Tal 2011: 1087). His work is relevant to our investigation because similar to quality of life the standard second lacks observable content that could be used to compare or test clock outcomes. The accuracy of atomic clocks like the accuracy of quality of life measures must be determined in lieu of comparisons to a gold standard.

The second lacks observable content because its definition is an ideal one: the second is defined as the duration of exactly 9,192,631,770 periods of the radiation corresponding to a hyperfine transition of cesium-133 in the ground state (see BIPM 2006). No actual cesium atom ever satisfies this definition; nor do we have a complete understanding of what it would take to satisfy it (see Tal 2011: 1087). The reasons why quality of life lacks observable content are somewhat different. Although there is no consensus among researchers on the definition of quality of life, most recognize that it entails "subjective" dimensions. In other words, quality of life is at least in part a matter of what respondents perceive and these perceptions are not observable.

Despite these differences, metrologists and quality of life researchers both face the challenge of "realizing" the referent of the constructs they aim to measure without having a complete understanding of what it takes to realize it. Metrologists must build clocks whose ticks approximate the theoretical definition of a second and quality of life researchers must construct questionnaires with items (questions) that elicit responses that approximate the definition that the instrument employs.

How do metrologists construct accurate clocks without appeal to a gold standard? As Tal (2011) discusses, metrologists realize the second by de-idealizing the theoretical definition in discrete steps (see ibid.: 1089). Each step identifies one way that a particular primary standard—or more accurately a cesium fountain—diverges from the theoretical ideal. The example that Tal provides is gravitational redshift (see ibid.: 1089–90). The definition of the standard second assumes that cesium is in a flat space-time, i.e. gravitational potential of zero. Primary standards, however, exist on earth where the gravitational potential is greater than zero. General relativity theory predicts that the cesium frequency will be redshifted, depending on the altitude of the laboratory where the particular primary standard is located. Redshifts thus indicate a kind of measurement bias not entirely different in principle from anxiety or response shift. The de-idealization process provides a magnitude for the predicted redshift and this correction is then added to the primary standard's outcome.

With every correction to the primary standard's outcome an estimate of uncertainty is also added. This estimate represents the range of values within which we are likely to find the true value of the correction. For instance, the magnitude of gravitational bias is calculated on the basis of a theoretical model of the earth's gravitational field and a measure of the primary standard's altitude. The measure of the primary standard's altitude is associated with a certain amount of uncertainty (as are all measurements), which affects the estimate of the redshift (see ibid.: 1090). The respective uncertainty associated with each correction yields a total uncertainty for each primary standard.

At this point it is helpful to acknowledge two different kinds of accuracy and their respective employment in the natural and social sciences. Epistemic accuracy is the closeness of agreement among values reasonably attributed to a quantity based on its measurement (see ibid.: 1085–6). For instance, the de-idealization of the second for a particular primary standard and its associated uncertainty represent the range of durations that reasonably represent the definition of the second. Metaphysical accuracy refers to the closeness of agreement between a measured value of a quantity and its true value (see ibid.: 1085). Estimates of the accuracy of physical quantities are most often expressed in terms of epistemic accuracy (see ibid.: 1086), but estimates of accuracy in the social sciences are often expressed in terms of metaphysical accuracy. We witnessed the reference to metaphysical accuracy in Oort et al.'s description of the problem that response shift presents: "observed changes in respondents' test scores may reflect something other than true changes in the attribute that we want to measure" (Oort et al. 2009: 1026).

How can metrologists have confidence that the uncertainty budgets of their primary standards are themselves epistemically accurate? Without a gold standard to compare the de-idealizations against how do we know when these models of de-idealization are adequate? Tal's response to this question is what he calls the Robustness Condition (RC) (see Tal 2011: 1090–1). RC specifies that the uncertainties ascribed to a primary standard's realizations are adequate if and only if (1) the outcomes of a primary standard converge on the outcomes from the other standards within the uncertainties ascribed to each clock and (2) the ascribed uncertainties are derived from appropriate theoretical and statistical models of each realization.

RC can also provide a measure of metaphysical accuracy (see ibid.: 1092–3). Estimates of uncertainty are from the metaphysical perspective the result of variables that confound the realization of the target construct. Thus, the convergence of multiple measures of the same construct within their ascribed uncertainties can provide a measure of how distant these realizations are with respect to the true value of the target.

Lessons for Quality of Life Research

What does this discussion of time tell us about quality of life? Two points become clear. First, it is not the lack of a gold standard that renders quality of life particularly difficult to measure. Recall that conventional wisdom in quality of life research tells us that quality of life is more difficult to measure than physical properties because it lacks (whereas physical properties do not) a gold standard. But this explanation of the difficulties of measuring quality of life is insufficient. Time, a physical measure of duration, also lacks a gold standard and yet it is a very successful measurement project.

As Tal discusses a large degree of the success of primary standards depends on RC, i.e. the outcomes from primary standards must converge within the uncertainties ascribed to each clock, and these uncertainties must be derived from appropriate theoretical and statistical models. Quality of life measures do not meet the criteria of RC; in fact, they are not in a position to attempt to meet them. I suggest that we should think about RC as directing our attention to some of the conditions for the possibility of measurement accuracy in lieu of a gold standard.[3] This may help us to see why quality of life instruments are unable to even approximate the conditions of RC.

RC suggests at least two conditions for its possibility. In order to meet the first criterion, measuring instruments must be modeled according to the theoretical and statistical assumptions regarding the construct one intends to measure, i.e. confounding variables and their associated corrections plus uncertainty are relative to the theoretical and statistical assumptions of the construct of interest. To meet RC's second criterion, the construct one aims to measure must be subject to significant theoretical and statistical assumptions, otherwise it is impossible to model the uncertainties associated with each realization of the construct.

This expansion of RC helps to locate the second lesson for quality of life: rich theoretical and statistical resources are necessary when measuring constructs that lack a gold standard. But quality of life lacks theoretical resources. While the lack of theory is not unacknowledged among researchers it is not given the attention it deserves. For instance, Hunt discusses this problem in her 1997 editorial and, more recently, in Donna Lamping's 2009 presidential address to the International Society for Quality of Life Research she cited the need for a theoretical framework as one of the three challenges facing this field (see McClimans 2010a: 225–6; Hunt 1997: 205–12). Nonetheless, researchers such as Hunt and Lamping often fail to link persuasively the lack of a theory of quality of life with the need for one. This failure is perhaps a historical artefact. Since its inception quality of life research has been policy driven, not theory driven. But without a theoretical framework how useful are these measures to policy?

Consider the constellation of problems that arise when we attempt to measure a construct without a theoretical representation of the measurement interaction. The most obvious problem is that without such a representation we lack a common definition of quality of life. As a result, we have hundreds of measures many of which operationalize quality of life differently. This difference affects the kinds of items (questions) in a particular instrument, the number of scales employed and the clinical significance of the outcomes. Imagine how confusing it would be if we had hundreds of clocks employing different definitions of the second; or hundreds of sphygmomanometers all of which measured blood pressure in different ways. Under these circumstances time and blood pressure would be rendered almost useless.

But the problems with quality of life are even more extensive. Without a theory of the measurement interaction quality of life also lacks a quantity concept, i.e. volume, mass, length, duration, pressure. Without a quantity concept we literally do not know what we are measuring. It is no surprise that in this context we have a great deal of difficulty interpreting the clinical significance of outcomes. What does a score of 10 on the Short-Form 36 9 (SF-36) mean? What does a change of 20 on the European Organization for Research and Treatment of Cancer Quality of Life Questionnaire (EORTC-QLQ-30) signify? Does it justify a change in a respondent's therapy? Does it indicate that a respondent's quality of life is better? Moreover, the problem with interpretation is related to problems with validity because if we do not know what the outcomes mean, then how can we determine if our measures assess what they aim to assess.

Finally, and with regard to the present discussion most importantly, as Tal's discussion of time makes clear, without a robust theory we do not have the grounds to justify taking some variable *a* to be a confounding variable. Recall that confounding variables are at least partially theory-relative, e.g. gravity is a confounding variable for primary standards relative to the theoretical definition of the second.

The persuasive differences between anxiety and response shift as possible confounding variables should now be obvious. While quality of life lacks a theoretical representation, blood pressure is situated within multiple overlapping theoretical frames: circulation, hypertension, gerontology, genetics, the qualitative work of doctors' observations of patients over time and the development of sphygmomanometers (see Mendlowitz 1979; Dana 1919; Hill and McQueen 1916).

Accordingly, although we still have much to learn about the causes of hypertension, we have good reason to believe that short-term anxiety of the kind typically experienced in the clinic is not a true cause of blood pressure. First, these short-term increases are theorized as being the result of our fight or flight system designed to enable us to optimize our survival when we perceive a threat. Second, by hypothesis when we experience anxiety our heart rate increases and blood is pumped throughout the body at a greater rate. This in turn should increase our systolic blood pressure for a limited time while generally leaving our diastolic pressure unchanged. Here we have a theory with a testable hypothesis that also provides a representation of the measurement interaction.

In contrast, response shift is taken to be a confounding variable in quality of life research, but the only reason we are given to believe this is a statistical one: response shifts violate measurement invariance. But this thin explanation presupposes that the quality of life construct is individuated from the response shift construct. Without a theory of quality of life researchers cannot explain why we should believe that these constructs are individuated. Without a theory the challenges that quality of life instruments encounter are not the same as those that sphygmomanometers face.

Is Quality of Life Immeasurable?

I have argued that the problems with quality of life measurement are not due to the lack of a gold standard, but rather the lack of a theory. But what of Hunt's claim that quality of life is a philosophic, specifically moral construct? Although in her editorial Hunt leaves it open whether quality of life can be measured, her emphasis on the dynamic and value-laden aspect of the construct has led others such as Gotay to assume that philosophic constructs are immeasurable ones (see Hunt 1997: 205–12). Although I am sympathetic to Hunt's emphasis on the dynamic and value-laden dimension of quality of life I think it is premature to suggest that it cannot be measured. Without a theoretical representation of the measurement interaction we cannot know if quality of life is "measurable"; indeed, we cannot even know what measurable means in this context. To the extent that we lack such a representation, we also lack understanding of quality of life's measurability.

Moreover, Hunt also says that because we do not know what quality of life is we cannot measure it (see ibid.: 209–10). In this she suggests a priority of questions: first we must know what something is and then we can ask how to measure it. But in the contemporary epistemology of measurement literature this priority is increasingly questioned (see Chang 2004; van Fraassen 2010). Bas van Fraassen (2010), for instance, writes in his book *Scientific Representation* that understanding what something is and what counts as a measure of it cannot be answered independently of one another (see ibid.: 116). Developing a theory of quality of life is thus intimately bound up with the development of the measuring instrument itself. This entanglement is why the theory that quality of life needs and lacks is one that provides a representation of the measurement interaction between the construct *and* the measuring instrument.

These questions of knowing and measuring are referred to in the measurement literature as the coordination problem: how do we link theoretical terms to empirical entities? If van Fraassen is correct about the entanglement of these questions then, to the extent that philosophy can help us to understand the measurement of quality of life, it can do so only to the extent that it provides a theoretical framework for quality of life and the questionnaires that attempt to realize it. Thus, a theory of quality of life should specify, for example, the relationship between quality of life and response shift. It should also attend to the logic of the questions used in the questionnaires. Moreover, an adequate theory must not only attend to each of these dimensions, but it must also be coherent with respect to both of them.

Van Fraassen writes that answering the entangled questions about the measurement construct and its measuring instrument has the form of a hermeneutic circle (see ibid.). This reference to hermeneutics means that we cannot understand the parts of a measurement project, i.e. the construct or the instrument, in isolation from the whole, i.e. the construct and its instrument. In other words, we cannot understand quality of life in isolation from what counts as a measure of it. But if we take van Fraassen seriously, then it may be a long time before we have an answer regarding the measurability of quality of life.

As Hans-Georg Gadamer cautions us, we cannot be quick to render a text or its analogue incomprehensible, that is, we cannot be quick to assume that a part and whole do not cohere, i.e. that quality of life is immeasurable (see Gadamer 2004: 268–72). In fact, we ought to assume provisionally that the parts and the whole will cohere, i.e. that quality of life is measureable. For Gadamer this presumption is required to ward against our natural aversion to change (see ibid.: 270–1). When parts and their wholes are finally understood, the understanding of the subject matter is often very different than that which we began. Change can be intimidating and difficult to process. If and when we do understand quality of life and the questionnaires as a coherent whole, then it is likely that our understanding of quality of life will be different than it is now. It is also likely that our understanding of measurement in this context will be different than it is now.

But change can also be good. As they now stand, quality of life measures have serious problems and yet they are used to contribute to the decision-making of very serious questions, questions about treatment, effectiveness, and quality of care. These questions are not only epistemically important, but also ethically important. As we saw earlier when we decide to individuate the constructs of quality of life and response shift, we say something not only about what quality of life *is*, but also something about the status of self-reports, i.e. that respondents have a distorted perception of reality, that their reports are the consequence of false consciousness.

Moreover, the need for a theoretical representation to justify taking response shift as a confounding variable becomes even more important when we recognize the role it plays in determining treatment effects. When we discount patients' adjustment to their circumstances, then the treatment effect generally increases, e.g. if your quality of life is still poor despite your adjustment, then clinicians, industry, and others can advocate for the usefulness of further treatment. Treatment that is successful in returning patients to their true quality of life will be considered effective. This example is not to suggest that clinicians or even industry are cynically taking advantage of response shift, but rather to note the systematic consequences that result from conceptualizing response shift as bias. The development of a theoretical representation of the quality of life measurement interaction is not just a theoretical concern; it is epistemic, it is ethical, and it is crucial.

Conclusion

Quality of life measures are known to suffer from a variety of problems: validity, interpretability, and ethics. While researchers generally agree about the existence of these problems they disagree about why they exist. I have argued that these difficulties are not the result of a lack of a gold standard. Instead I locate these difficulties in the lack of a theoretical representation of the measurement interaction between quality of life and the questionnaire that attempt to measure it. To those who suggest that quality of life is a philosophical construct and cannot be measured, I argue that this position is premature. Without a representation of the measurement interaction we cannot know what quality of life is or what counts as measuring it. Indeed, if we follow van Fraassen in thinking of measurement as a hermeneutic problem, then we ought to presume that quality of life *is* measurable.

Notes

1 Carolyn Gotay is Professor and Canadian Cancer Society Chair in Cancer Primary Prevention at the University of British Columbia. In 2011 she received ISOQoL's President's Award, an honor given annually to commend exceptional efforts to advance quality of life through health education, promotion, research or policy initiatives.

2 Ableism refers to the unquestioned assumption that disabilities diminish quality of life. If one holds this assumption, then when those with disabilities report having a good quality of life these reports are understood as reflecting bias.

3 Insofar as RC provides the criteria for metaphysical accuracy these conditions also help to specify conditions of validity, i.e. how well does an instrument empirically describe the theoretical entity it aims to measure?

References

BIPM (Bureau International des Poids et Measures). 2006. *The International System of Units (SI)*. 8th edn Sèvres: BIPM, http://www.bipm.org/en/si/si_brochure/, accessed March 15, 2016.

Bridgman, P. W. 1980. *The Logic of Modern Physics*. New York: Arno Press.

Chang, H. 2004. *Inventing Temperature: Measurement and Scientific Progress*. Oxford: Oxford University Press.

Dana, H. W. 1919. Theories Regarding Blood Pressure. *Journal of the American Medical Association 72:20*: 1432–4.

Darzi, Lord of Denham KBE 2008. High Quality Care for All: NHS Next Stage Review Final Report. http://webarchive.nationalarchives.gov.uk/20130107105354/ http://www.dh.gov.uk/en/Publicationsandstatistics/Publications/Publications PolicyAndGuidance/DH_085825, accessed February 09, 2014.

Editorial Quality of Life and Clinical Trials. *The Lancet 346(1995)*: 1–2.

Gadamer, H.-G. 2004. *Truth and Method (Continuum Impacts)*, 2nd revised edition, London: Continuum.

Hill, L. and McQueen, J. M. 1916. The Theory of Blood Pressure Measurement. *British Medical Journal* 1: 875–8.

Hobart, J. C., Cano, S. J., Zajicek, J. P., and Thompson, A. J. 2007. Rating Scales as Outcome Measures for Clinical Trials in Neurology: Problems, Solutions, and Recommendations. *Lancet Neurology* 6: 1094–105.

Hunt, S. M. 1997. The Problem of Quality of Life. *Quality of Life Research: an International Journal of Quality of Life Aspects of Treatment, Care and Rehabilitation* 6: 205–12.

Kerr, E. A., Zikmund-Fisher, B. J., Klamerus, M. L., Subramanian, U., Hogan, M. M., and Hofer, T. P. 2008. The Role of Clinical Uncertainty in Treatment Decisions for Diabetic Patients with Uncontrolled Blood Pressure. *Annals of Internal Medicine 148:10*: 717–27.

McClimans, L., Bickenbach, J., Westerman, M., Carlson, L., Wasserman, D., and Schwartz, C. 2013. Philosophical Perspectives on Response Shift. *Quality of Life Research: an International Journal of Quality of Life Aspects of Treatment, Care and Rehabilitation* 22:7: 1871–8.

McClimans, L. 2011. Interpretability, Validity, and the Minimum Important Difference. *Theoretical Medicine and Bioethics* 32: 389–401.

McClimans, L. 2010a. A Theoretical Framework for Patient-reported Outcome Measures. *Theoretical Medicine and Bioethics* 31: 225–40.

McClimans, L. 2010b. Towards Self-determination in Quality of Life Research: a Dialogic Approach. *Medicine, Health Care, and Philosophy* 13: 67–76.

Mendlowitz, M. 1979. Some Theories of Hypertension: Fact and Fancy. *Hypertension* 1: 435–41.

Oort, F. J., Visser, M. R. M., and Sprangers, M. A. G. 2009. Formal Definitions of Measurement Bias and Explanation Bias Clarify Measurement and Conceptual Perspectives on Response Shift. *Journal of Clinical Epidemiology* 62:*11*: 1126–37.

Powers, B. J., Olsen, M. K., Smith, V. A., Woolson, R. F., Bosworth, H. B, and Oddone, E. Z. 2011. Measuring Blood Pressure for Decision Making and Quality Reporting: Where and How Many Measures. *Annals of Internal Medicine* 154:*12*: 781–8.

Salzberger, T. 2013. Attempting Measurement of Psychological Attributes. *Frontiers in Psychology* 4:*75*: 2–3.

Schwartz, C. E. and Sprangers, M. A. 1999. Methodological Approaches for Assessing Response Shift in Longitudinal Health-Related Quality-of-Life Research. *Social Science & Medicine* 48:*11*: 1531–48.

Streiner, D. L. and Norman, G. R. 2008. *Health Measurement Scales: A Practical Guide to Their Development and Use.* Oxford, New York: Oxford University Press.

Tal, E. 2011. How Accurate Is the Standard Second? *Philosophy of Science* 78:*5*: 1082–96.

Van Fraassen, B. C. 2010. *Scientific Representation: Paradoxes of Perspective.* Oxford: Clarendon Press.

Wasserman, D., Wachbroit, R., and Bickenbach, J. 2005. *Quality of Life and Human Difference: Genetic Testing, Health Care, and Disability.* Cambridge: Cambridge University Press.

11 Measuring Intelligence Effectively: Psychometrics from a Philosophy of Technology Perspective

Andreas Kaminski

The history of psychometrics has one constant: the enthusiastic development of means for measuring various kinds of mental performance (to which I will refer simply as measurement of mind) is accompanied by a shadow-like critique that is no less forceful. In the mid-19th century, when Gustav Theodor Fechner—following on from Johann Friedrich Herbart and Ernst Heinrich Weber—presented the idea of psychophysics with great verve, he often met with vehement rejection alongside widespread approval. Critical objections have often been raised on principle, such as that in the domain of the psyche only intensive quantities exist which cannot be measured (see Kant 1956 [1787]: B207–218). Another point of critique has been that there is no adequate psychological measure for the psyche. When research on intelligence began around 1900: it very quickly became an area of international research which was rapidly applied in schools, the military, corporations, and psychiatry. To this day, intelligence research has been vehemently criticized, partly because of its alleged dependency on language and culture. Likewise, the measurement of competence, which developed in the 1970s, has also been accompanied by its Siamese twin, critique of the measurement of competence. Most of these critiques originate from the philosophy of science. In the following, I would like to show that this approach is not sufficient. It needs to be supplemented by arguments from the philosophy of technology.

In order to explain why I believe this to be the case, I begin by presenting two theses that serve to illuminate the connection between psychology and the philosophy of technology.

The first thesis is: Psychology has practical and not just epistemological implications. The American philosopher Richard Rorty summed up the debate around psychologism by noting that "the rise of empirical psychology had raised the question 'What do we need to know about knowledge which psychology cannot tell us?'" (Rorty 1979: 165). While Rorty's question manages pointedly to expose the heart of the debate around psychologism, the way he formulates it simultaneously reveals an obvious blind spot. Psychology and psychometrics have been understood as challenges to the theory of cognition and the philosophy of science. But is this primarily what they are? I have my doubts, and I would like to justify them using the case of psychological examination techniques. The debate about psychometrics has revolved particularly around the epistemological question of whether it is *possible* to measure mental performance. Beyond this epistemological question, however, lies a further issue, which I elucidate in the following.

Psychometrics is effective inasmuch as it influences the ways in which people relate to themselves (their self-relation), and it does so regardless of the answer to the question of its feasibility or ability to produce reliable knowledge. Psychometrics may be an impossible science epistemologically, but it is nonetheless an effective technology because it changes a person's self-relation. This is because the benchmark used in psychometrics is not external. If I were to measure the length of an object by placing a measuring rod next to it, my act of measuring does not change the object. If I measure the intelligence of a person, however, this benchmark does not remain external to that person. Why?

The difference is not that, in the one case, the influence of the measuring tool on the object remains negligible, whereas in the other case—the measurement of intelligence—the measuring tool is of practical relevance. In the second case, it is not a matter of *influence*, but rather a matter of the *change in the subject's self-relation*, which may result in the individual forming a concept of his or her self from the point of view of the benchmark. This in turn may lead them to adjust their behavior in order to conform to the benchmark. Thus, the problem with any psychological examination technique lies in the fact that—to use the analogy of the measurement of length—the subject (the person) begins to "stretch" maybe as soon as the psychometric benchmark is applied, because it knows that it is being measured.

The case of psychometrics is thus highly suitable as an example to point out what seems to me to be a characteristic shared by psychology and by the debates surrounding it: it is concerned not only with implications in the fields of scientific theory or the theory of cognition but also with social and technical implications.

This is the second thesis: The success of psychology is due not to its theoretical principles but rather to its potential for technical application. The rise of psychology is not correlated with advances made in basic knowledge or fundamental understanding but with the fact that it opened up a number of fields of application for itself after 1900. The conviction that psychology offers solutions to "practical problems"—the "tasks of culture"—is based on the notion that it is capable of providing answers to certain questions in a technical form, as it were, specifically in the form of measurements. However, the models used in psychology (such as cognitive load theory) represent the psyche in a manner analogous to technology. Where psychologists—far away from psychoanalysis—have been engaged in practical work, it is certainly not their underlying theories that have been most relevant. The lasting contribution of William Stern, for example, has not been his theory of intelligence or his coining of the concept of psychotechnics or even his theory of personality;

what has attracted attention and gained widespread acceptance is his notion of an intelligence quotient, or "IQ." To put the point somewhat starkly: in contexts where it has had practical relevance, psychology in the 20th century has always also been "psychotechnics".

My contribution is divided into three parts. In the first part, I will discuss three key arguments in the critique of psychometrics. My purpose is to show that the critique of psychometrics has an underlying blind spot. The different positions I discuss share a common perspective which is so self-evident to them that they fail to recognize a major missing factor. The question they all pose is whether it is *possible* to measure the mind. In doing so, however, they fail to take account of the *effectiveness* of psychometrics, that is, the fact that psychometrics changes not only social relations but also self-relations.

In the second part, I will attempt to show what the reasons are for the effectiveness of psychometrics by comparing psychometric and classical forms of measurement. My thesis is that there is one important difference in psychometrics: self-relations are influenced by psychometric measurement techniques, because subjects behave in relation to the scale which is used to measure them.

In the third and final part, and using the example of the Lynn-Flynn effect, I will try to define the consequences of this behavior of subjects in respect to scale. This part will demonstrate the extent to which the philosophy of subjectivity is relevant to psychometrics.

The Critique of Psychometrics—and its Blind Spot

It is possible to identify three distinct positions within the critical epistemological debates surrounding psychometrics. They offer valuable analyses of the problems entailed in acquiring knowledge by measuring mental performance and they are also helpful when it comes to evaluating the legitimate and illegitimate claims made by psychometrics. However, these three critical positions share a common perspective that serves to render a key issue invisible. In the following I seek to make this blind spot visible by looking at each position in turn.

The first critical position assumes that measuring mental performance is *possible in principle*, even if in practice the methods for doing so require improvement. Denny Borsboom, for example, assumes that psychometrics is a valid approach in principle but also points out that its weakness lies in the lack of appropriate theories to address the object being measured. Borsboom sees considerable effort being expended upon the development of methods and technologies of measurement, yet without any corresponding development of a substantial theory of the measured object. Without such a theory, says Borsboom, the validity of the measurements cannot be verified (see Borsboom 2005).

The second critical position argues that it has not yet been established *whether* the measurement of mental properties and performance is *possible*.

This might or might not be the case—it has not yet been determined. Joel Michell has developed this position in a series of publications. Michell frames the question of the feasibility of psychometrics by noting that it needs to accomplish two tasks, a scientific one and a methodical and technical one. The scientific task consists in showing that the prerequisite for measurement—namely the quantitative nature of the object—is satisfied. In order to establish criteria that determine whether or not an object is quantitative, Michell turns to the work of mathematician Otto Hölder (1901) who defined the "axioms of quantity" (see Hölder 1901). Michell assumes that the proof of an object having a quantitative nature can only be provided empirically, and he develops methods for doing this. His criticism of psychometrics is that it has never taken on this first scientific task. Instead of demonstrating that its object is quantitative, psychometrics has simply acted as if it were so. It has largely devoted its efforts to the development and refinement of methods, none of which would be of any use if applied to an a priori inadmissible object. In particular, Michell critically notes that psychometrics does not recognize that the question of whether or not psychometrics itself is possible remains unresolved (see Michell 1997; 2000; 2005). In other words, the feasibility of psychometrics has yet to be determined.

The third position is based on the assumption that measuring mental properties or performance is (probably) *impossible*. Even if a proof of the quantitative nature of the mind were given—as required by Michell—psychometrics would be an impossible science because the object, the psyche, does not satisfy other pre-requisites for measurement. A critique of psychometrics along these lines has been put forward by Trendler (see Trendler 2009). In order to be measurable, an object must be capable of being clearly located within a simplified causal nexus: apart from a few causal relations, all others would have to be excluded. In addition, it would have to be possible to manipulate rather precisely the remaining causal relations. Within the natural sciences devices and technical environments have been created which satisfy both conditions. The same has not been true regarding the psyche. Neither causal reduction nor causal manipulation is given to a sufficient extent, and there is no prospect of any change in this situation. For this reason, psychometrics is not feasible as a science.[1]

What do these three critical positions have in common? The first assumes that while it is theoretically possible to gain information about the mind by means of measurement, methodological improvements are required to do so. The second deems it to be undetermined whether measurement of the mind is possible in principle, while the third denies this possibility on practical grounds. These three strands of criticism are focused upon the possibility or impossibility of psychometrics. What is being addressed is the relationship between measurement and mind. All three positions examine whether this relationship is suited to the task. Is the object suitable for measurement (positions 2 and 3)? Is the measurement theory suited to producing an adequate measurement of the object (position 1)?

These approaches seem so self-evident that it is indeed difficult to identify what they do not take into account. As valuable as the critiques of the claims of psychometrics are, they disregard an important point, namely, that psychometrics is *effective*, no matter whether—epistemologically speaking—it is feasible or not. By focusing on the issue of whether or not psychometrics is *feasible* we have lost sight of the fact that it is *effective* (even if its claims are epistemologically inadmissible). The effectiveness of psychometrics can be observed in the way it changes social self-relations. People appreciate themselves differently when they see themselves reflected in the mirror of intelligence tests or measurements of competence. They regulate their behavior in different ways, aligning themselves with the scales of the measurement being conducted.

In order better to understand this effectiveness, we need to proceed to the next issue: what distinguishes psychometrics from classical methods of scientific measurement?

An Object that Behaves According to the Scale Applied to It

Let us turn to the question of how psychometrics differs from classical measurement practices. By way of example, we can compare the measurement of length or temperature with that of intelligence. The most familiar attempts at establishing the differences can be traced to the differentiation of intensive and extensive properties, as described by Kant in his *Critique of Pure Reason* (see Kant 1956 [1787]: B207–218). In order to contest the measurability of mental properties, let us examine the attempts at differentiation which (more or less clearly) followed Kant's line of argument.

Mental properties are not additive

This approach dates from Norman Campbell's theory of fundamental measurement. Campbell assumed that criteria can be stated for objects that are measurable (see Campbell 1920: 267–94). Among these criteria is the requirement that the objects' properties must be transitive, asymmetrical and, above all, additive. Objects' properties are additive if empirical operations of addition can be stated for them. Only then can corresponding mathematical operations be conducted with these properties. Concerning weights and lengths, this means that weights can be added to the scale pan or lengths can be increased by placing another object next to the original one: these are empirical operations of addition. As far as mental properties are concerned, no such operation can be readily identified. The problem with this criterion is that Campbell mentions fundamental as well as derived measurements. Density, for example, does not allow for an immediate empirical operation of addition. However, it is measurable by derivation, through the use of other properties which themselves are additive. In principle, such procedures could be found for mental properties, as psychophysics has attempted to do.

By doing this however, the proposal would lead back to the debate about whether mental properties are quantitative or not. This is why the criterion is inadequate: it merely leads back to the unanswered question asked by Michell (see Michell 1997).

There is no mental measure

Could the fact that there is no mental measure (or that no such measure can be set as a stable standard) constitute a distinctive feature? If this were the case, measurement of the mind would be different from that of nature. While neither has any pre-ordained measures, nature provides us with measures which, due to their far greater stability and commensurability, are much more suited as scales of measurement. As a way out of this, Fechner devised a very clever notion of using those differences in perception which are just noticeable as a scale. Quite apart from the controversy surrounding psychophysics, though, a similar approach seems impossible for other mental phenomena (other than sensation).

However, the history of measurement practices shows that other classical measurement methods posed similar problems in their early stage of development. In his detailed account of how temperature came to be a scientific concept, Hasok Chang has shown how difficult it was to find "fixed points" in temperature measurement and to define them in a way that would allow them to be applied in practice (see Chang 2004). When does water boil? When does it freeze? Those seemingly natural "fixed points" seem to the naked eye to be more like "ranges" rather than points and are therefore quite unsuitable to fix the calibration of a thermometer, as Chang has shown. Similarly, intelligence does not display an immediate, quasi-natural reference point, either. In any case, the mental measure of 1 IQ has a different status than 1 m. But here, too, statistical procedures have made it possible to find "fixed points," namely, by generating average points on a distribution curve (see Terman 1916).

Failure and alternative

These obvious attempts at differentiating between classical and mental measurement therefore fail. Indeed, the attempts at differentiation suggest that one might conceptualize subjectivity as something that is romantically indeterminable, as it were. The following proposed means of differentiating between classical and mental measurement does not proceed on the assumption of such a romantically inscrutable subject.

In psychometrics—unlike in classical forms of measurement—the object exists in active relation to the benchmark applied. This is not the case with the measurement of so-called external nature. No body that has a ruler laid against it adapts its size to the scale—i.e. by becoming or seeming to become larger. No temperature that is measured adapts its heat to the scale of the thermometer applied to it. People who have psychometric measurement techniques applied to them, however, will generally adapt themselves to the benchmark, that is to say, to the test criterion. People who are subjected to an intelligence test will by and large seek to behave in such a way that the test result certifies high intelligence. In general, when someone is subjected to a competence test, they will attempt to appear at their most competent. Comparing this to the measurement of length, this would be the equivalent of the object being measured stretching itself to appear as long as possible.

A general difference between psychometric testing techniques and classical measurement techniques thus becomes plain: in psychometrics the criterion is not external to the subject. Instead, the subject relates to the criterion which is applied to it. The classical relation is that between measurement and object. This is the point on which psychometrics's classical critical positions focus, as seen above. They ask whether this relation is possible and whether it has been adequately conceptualized. However, this is where it becomes apparent that the relation between measurement and object occurs in the object—in psychometrics, the subject—itself; it is reflected by it and affects its behavior. Within the context of psychometrics, the subject behaves relative to this relationship, and therein lies the effectiveness of psychometrics, which exists regardless of its epistemological possibility.

The effectiveness of psychometrics is by no means restricted to the actual test event. In order to understand this, it is important to realize what a benchmark is when seen in the context of psychometrics. It is more than merely a scale; it is also a dimension of subjectivity, as a benchmark indicates what subjects are capable of being (or becoming). Intelligence tests not only determine a person's general ability to perform certain mental tasks; in order to do this, they also need to state what subjects may be (i.e. "intelligent" or, to take other standards, "weak-/strong-willed," "extroverted," "able to deal with conflict"). What subjects can be, what defines their subjectivity, what their subjectivity consists of, changes with the appearance of intelligence tests. At an elementary level, the self-conception of those who ask themselves "who am I?" and "who can I be?" changes when their intelligence is measured.

This comes to the fore with intelligence tests in particular, given that even today such tests are associated with scholarly and professional achievement. Intelligence tests are regarded as predictive tools—they purport to predict the future of a person in an important area of their personality. Measurement procedures also change people's self-conception in other areas of cognition. Thus personality and competence testing define a grid of potential subjectivity. They construct a "space of possibility" that shows how subjects can be—and they assign an individual a certain location within this space.

Alongside the issue of self-conception, psychometrics acts on another level, namely, that of *work done on the self*. The benchmark not only creates an understanding of what and how subjects can be but also serves as a prompt for self-control and self-formation. The scale of the benchmark is a scale of values. It allows for both downward and, especially, upward orientation. It gives feedback on where a person is and whether progress is being made. Practice, self-control and self-formation are aspects of the effectiveness of psychometric processes.

For the reasons just outlined, then, it is not sufficient merely to ask whether measurement is possible. Rather, the issue that requires additional attention is that it is effective. Regardless of whether or not the mind is quantitatively constituted and regardless of whether or not it is accessible to measurement from an epistemological perspective: since the object being measured exists in active relation to the means of measurement, the act of measurement can be described as effective because of the behavior it induces. In other words, the benchmark is not external to the subject but can rather be acquired by the subject as a powerful social benchmark. The consequence of this is that our society and our self-reference (because an ambitious notion of *self* can only be conceived of in the context of social relations with others) changes with psychometrics—and with the introduction of new psychometric methods of testing. These change people's self-conceptions and the way they work on (act upon) themselves. To name two of the breaks within this psychometric history of subjectivity, humans became intelligent around 1900 and became competent around 1970!

Traces of Psychometrics: the Lynn-Flynn Effect as a Change in Ways of Thinking

In the first part of this chapter I showed that, in asking first and foremost whether psychometrics is possible, scientific theory loses sight of the reality and effectiveness of the measurement practices of the mind. In the second part I established this effectiveness by showing that the object—the subject—behaves relative to the benchmark applied to it. With psychological measurements, the self-conception and the work a subject does on herself change. These two observations may create the impression of two perspectives that co-exist side by side without having much to do with each other: the epistemological perspective on the one side and the perspective inspired by the philosophy of subjectivity on the other. I do not believe this to be the case, however. The discussion in the second part of the chapter regarding the reflexivity prompted by a benchmark has consequences for epistemological issues. In this third part I would like to offer, all too briefly, a few indications of what these consequences are.

The so-called Lynn-Flynn effect belongs to the history of the measurement of intelligence. It was first observed by psychologist Richard Lynn and was made more widely known by political scientist James R. Flynn in the 1980s. This effect consists in a significant rise in the level of intelligence recorded over the course of a few decades. The background is as follows: intelligence tests have to be standardized (that is, re-calibrated) repeatedly. This is done by selecting a representative sample of test subjects (the standardization sample) in order to establish new measurement norms. The standardization is done using the sample average, which is defined as being 100 points. When this process of standardization is repeated, changes in the level of intelligence of the standardization samples become invisible. This was the starting point for Flynn's idea. He calculated the scores for later test samples using the earliest standardization, which went back to the year 1932 in his data set. The result was that all the later samples showed a continuous, relatively linear, and equally observable increase in the level of intelligence for all groups within the normal distribution range when they were calculated uniformly using the standardization from 1932. Flynn examined test samples from 1932 to 1978 in this way and observed an increase in the average by about 13.8 points (see Flynn 1984).

Various explanations have been offered for this effect, from changes in nutrition through a change in environmental influences (media) to improvements in medical care. Some researchers have commented that intellectual and abstract ways of thinking have become more a part of everyday life—from schools to work environments—than was the case 100 years ago. Without wishing to ally myself with any single monocausal explanation, any one of these inferred explanations would provide a close match for the arguments presented in the second part of this chapter. Intelligence tests apply a different benchmark for measurement than that of mere memorization assignments: subjects adjust their behavior relative to an intelligence benchmark—which is simultaneously a value-based one and one which holds out the prospect of scholarly, professional and financial success. Work on the self and self-control relative to this benchmark, lessons at school and exercises at home have changed. Given that intelligence is a (not merely natural) disposition and that every disposition must be developed by work, it may not be so surprising after all that subjects' ways of thinking have changed over time; after all, intelligence tests have had and still have a good reputation. Broadly speaking, the subjects of psychometric testing are not just thinking about different *content*, they are thinking in a different *way* than they did 200 years ago.

Note

1 For objections to this argument see Markus and Borsboom (2012).

References

Borsboom, D. 2005. *Measuring the Mind. Conceptual Issues in Contemporary Psychometrics*. Cambridge, New York: Cambridge University Press.

Campbell, N. R. 1920. *Physics. The Elements*. University of Michigan: University Press.

Chang, H. 2004. *Inventing Temperature. Measurement and Scientific Progress*. Oxford, New York: Oxford University Press.

Flynn, J. R. 1984. The Mean IQ of Americans: Massive Gains 1932 to 1978. *Psychological Bulletin* 95:*1*: 29–51.

Hölder O. 1901. Die Axiome der Quantität und die Lehre vom Mass. *Berichte über die Verhandlungen der Königlich Sächsischen Gesellschaft der Wissenschaften zu Leipzig, Mathematisch-Physische Classe* 53:*1*: 1–64.

Kant, I. 1956 [1787]. *Kritik der reinen Vernunft*. Hamburg: Meiner.

Markus, K. A. and Borsboom, D. 2012. The Cat Came Back: Evaluating Arguments Against Psychological Measurement. *Theory & Psychology* 22:*4*: 452–66.

Michell, J. 2005. *Measurement in Psychology. A Critical History of a Methodological Concept*. Cambridge: Cambridge University Press.

Michell, J. 2000. Normal Science, Pathological Science and Psychometrics. *Theory & Psychology* 10:*5*: 639–67.

Michell, J. 1997. Quantitative Science and the Definition of Measurement in Psychology. *British Journal of Psychology* 88:*3*: 355–83.

Rorty, R. 1979. *Philosophy and the Mirror of Nature*. Princeton: Princeton University Press.

Terman, L. M. 1916. *The Measurement of Intelligence. An Explanation of and a Complete Guide for the Use of the Stanford Revision and Extension of the Binet-Simon Intelligence Scale*. Cambridge: The Riverside Press.

Trendler, G. 2009. Measurement Theory, Psychology and the Revolution That Cannot Happen, *Theory & Psychology* 19:*5*: 579–99.

12 The Klein Sexual Orientation Grid and the Measurement of Human Sexuality

Donna J. Drucker

As sex research developed into a scientific field in the late 19th century, researchers have struggled with how something as mysterious, ephemeral or to use the word that many other scholars have—fluid—could be ordered for scientific study. Sex researchers from the 1920s onward, including Robert Latou Dickinson, Katharine Bement Davis, and Gilbert V. Hamilton, experimented with quantitative measurements for their work to complement the qualitative (see Dickinson and Beam 1932 and 1934; Hamilton 1929; Davis 1929). Examining quantitative forms of sexual measurement illuminates not only what such measures have been for, but also whom they have been for—how effective (or not) they are in developing scientific thought, and also how useful they are for public and individual understanding of sexuality. This essay focuses on one of the best known quantitative measurements, the Klein Sexual Orientation Grid (aka KSOG, or Klein Grid)—its theoretical and historical development, its use in scientific study, its transference to the Internet, and its use in personal discovery, and what it does and does not show about human sexuality. Overall, the Klein Grid is a tool best designed for researchers and professional psychologists working with clients, although it, along with other tools, remains popular with and relevant to individuals seeking instruments for personal discovery.

Fritz Klein, the creator of the grid that bears his name, was born in Vienna in 1932 but moved to New York as a small child. He returned to Europe as an adult to attain his doctorate in medicine at the University of Bern, Switzerland, and later completed a master's of business administration degree at Columbia University in New York City. He divided his adult life as a board-certified psychiatrist between New York and San Diego, California. As part of his involvement in the sexual revolution of the 1970s, he founded "The Bisexual Forum" in those two cities, where bisexuals met to discuss common problems and to create community. He authored *The Bisexual Option* in 1978 (reprinted 1993). He was also the founder and first editor of the academic *Journal of Bisexuality*, and remained involved in research and education concerning bisexuality until his death in 2006.[1]

While the scientific discourse about bisexuality began with Sigmund Freud and Freud's term "psychological bisexuality," Klein's work initiated a new

Klein Sexual Orientation Grid	Past	Present	Ideal Future
A. Sexual Attraction (To whom are you attracted sexually?)			
B. Sexual Behavior (With whom have you actually had sex?)			
C. Sexual Fantasies (Whom are your sexual fantasies about?)(They may occur during masturbation, daydreaming, as part of real life, or purely in your imagination.)			
D. Emotional preference (Do you love and like only members of the same sex, both sexes or other sex?)			
E. Social Preference (Do you like to socialize with member of same sex, both sexes or other sex?)			
F. Lifestyle Preference (Do you basically live a straight lifestyle, mixed lifestyle or gay lifestyle?)			
G. Self-Identity (How do you think of Yourself?)*			
Copyright © 1978 American Institute of Bisexuality, Inc.			

Figure 12.1 The Klein Sexual Orientation Grid (KSOG) by Fritz Klein. Reproduced with the permission of American Institute of Bisexuality (www.bisexual.org).

iteration of that conversation based in the women's, gay, and lesbian and sexual liberation movements in the 1970s (see Gooß 2008). Klein's inspiration for the grid derived from his dissatisfaction with the state of scientific research on bisexuality specifically and on sexuality generally at the time. Klein found in his practice that the concept of "bisexuality" (not to mention the concepts of "heterosexuality" and "homosexuality") had no fixed meaning, and that individuals with widely varying combinations of desires, orientations, and behaviors used the word "bisexual" to describe themselves. As Klein himself put it, "No matter what definitions we gave to the three categories of sexual orientation, limiting it to only these three possibilities did not do justice to what these people knew about themselves and others" (Klein 1978: 15). He also found contemporary models and research tools that aimed to capture the complexity of sexuality to be inadequate. So he decided to create a modified version of the Kinsey Scale, which the Indiana-based entomologist-turned-sex-researcher Alfred C. Kinsey first drew in 1940 and then published in *Sexual Behavior in the Human Male* (aka *Male* volume) in 1948. Klein stated bluntly in *Bisexual Option* that: "No matter what sexual orientation a person has, he or she lives on a continuum" (ibid.: 8).

Kinsey's 0–6, or heterosexuality to homosexuality scale, marked desire, fantasy, and behavior in a person's sexual history onto a 7-point scale (see Kinsey et al. 1948: 638; Drucker 2010). (Asexuality was a separate element from the 0–6 scale, and the authors discussed it in more depth in the 1953 volume *Sexual Behavior in the Human Female,* aka *Female* volume (see Kinsey et al. 1953).) In the *Male* and *Female* volumes (together known as the Kinsey Reports), Kinsey or a member of his interview team selected the single-digit number for an interviewee that they felt best matched the interviewee's feelings and experiences. A "Kinsey 0" was a person who had only heterosexual inclinations, a "Kinsey 6" was a person who had only homosexual inclinations and the numbers 1–5 measured stood for proportionate degrees of desire, fantasy, and behavior in between. Although Kinsey and his research team assigned a single round number to each of their 18,000 interviewees for the purpose of statistical analysis, Kinsey knew that there were infinite variations for an individual's place along the scale: "Males do not represent two discrete populations, heterosexual and homosexual. The world is not to be divided into sheep and goats. Not all things are black nor all things white. It is a fundamental of taxonomy that nature rarely deals with discrete categories" (Kinsey et al. 1948: 638). However, without Kinsey's textual explanation, readers who only saw the Kinsey Scale reprinted in magazines or newspapers would likely have thought that the single numbers were the only options for an individual's placement on the scale. Another problem with the Kinsey Scale was that it did not measure a person's subjective sense of sexual identity or orientation.

The lack of identity and orientation considerations on the Kinsey Scale was particularly troubling to Klein as he prepared his own instrument of measurement. Klein had found prejudice against self-identified bisexuals from both homosexuals and heterosexuals through his bisexual activism. So he also wanted to raise awareness of people with bisexual identities and to give them opportunities to establish broader political and social awareness for themselves, himself included.[2] Klein first published the Grid in his advice book *Bisexual Option* and then in the pornographic magazine *Forum* (later *Penthouse Forum*) in 1980.[3] He printed a copy of the Grid in the magazine, asked *Forum* readers to fill out the Grid and then requested that they return it to him. He received 584 responses and published an article in 1985 with two colleagues describing them. The 1985 article brought the Grid to broader scientific attention. In Klein's own words, the Grid "was developed to measure a person's sexual orientation as a dynamic multi-variable process. The Grid was designed to extend the scope of the Kinsey Scale by including attraction, behavior, fantasy, social and emotional preference, self-identification and lifestyle. These characteristics are also measured in the past, present, and as an ideal" (Klein et al. 1985: 38).

For the purpose of Klein's study, "the past" was an individual's life up to 12 months ago, and "the present" was the most recent 12 months, and "the ideal" was how individual would eventually like to live. The scale had 21 elements total. An individual completing the scale would rate the first five elements on a scale of 1 to 7: (1) other sex only (2) other sex mostly (3) other sex somewhat more (4) both sexes (5) same sex somewhat more (6) same sex mostly and (7) same sex only. For the last two elements, an individual would rate the points on a different scale: (1) heterosexual only (2) heterosexual mostly (3) heterosexual somewhat more (4) hetero/gay-lesbian equally (5) gay-lesbian somewhat more (6) gay-lesbian mostly and (7) gay-lesbian only. The differences among the ratings "mostly" and "somewhat more" are left to the individual, which introduces subjective judgment is also a recurrent problem with Likert Scales (gradations of agreement and disagreement with a statement) (see Likert 1932–33). The questions were framed as follows:

(1) To whom are you attracted?
(2) With whom have you actually had sex?
(3) Whom are your fantasies (during masturbation, daydreaming, real life or imagination) about?
(4) Emotions influence, if not define, the actual physical act of love. Do you love and like only members of the same sex, only members of the other sex or members of both sexes?
(5) Social preference is closely allied with but often different from emotional preference. With members of which sex do you socialize?
(6) What is the sexual identity of the people with whom you socialize? How do you think of yourself?
(7) Some people describe their relationship to the rest of society differently than their personal sexual identity. For instance, a woman may have a *heterosexual* sexual identity, but a *lesbian* political identity. How do you think of yourself politically?[4]

The most inventive aspect of the Klein Grid as compared to the Kinsey Scale is the addition of the ideal set of circumstances through which one could live one's sexual, social and political life. Researchers and professional sex therapists or educators who used the Klein Grid could then compare subjects' actual lives to their ideal lives, to see the distance, or lack thereof, between them. The future or ideal element was especially useful for clinicians, who could then help clients work toward the integration of their sexual identities into their overall senses of self and future goal-setting. The Klein Grid, along with the Kinsey Scale, illustrates that "how we attempt to 'measure' sexual orientation or desire not only determines the proportions of the population we assume to be bisexual, heterosexual and homosexual, but also reflects the different ways we conceptualize sexuality in general, and bisexuality in particular" (Udis-Kessler 1999: 49). Even if a clinician did not agree with Klein's broader concept of bisexuality, one could use the Grid to gain insight into the identity issues facing a client.

However, the Klein Grid is problematic for multiple reasons, some of which speak to the historical context of the scale's production. For example, the idea of a difference between a political sexual identity and a personal sexual identity—as in the political lesbianism in the example—is now a little-used identity framework, more popular in the second-wave feminist and sexual politics of the 1970s.[5] Furthermore, Klein's questions about emotional, social and lifestyle preference were problematic for test-takers and other research-ers, as many saw that having opposite-sex friends and acquaintances had little or no impact on their sexual identity and orientation. Also, as Kinsey Institute researchers showed in the 1990s and the 2000s, the phrase "had sex" can have a broad variety of meanings, and thus test-takers' answers to these questions would not be consistent (see Sanders and Reinisch 1999; Sanders et al. 2010). Neither does the Klein Grid ask specifically about orgasm for either the test taker or that person's partners. It also treats each of its com-ponents on the same plane, as if each component weighs equally in deciding sexual orientation. As one scholar summarized deficiencies in the Klein Grid:

> The KSOG itself is not exhaustive, and arguably excludes factors which, although perhaps more difficult to "measure" on scales like the KSOG, nevertheless form an important part of sexual subjectivity. For example, one's perception of one's own body and erogenous zones—one's sexual or erotic body image—and the ways one perceives the sexual body of one's partner(s) will both play an important part in one's sexuality.
>
> (Storr 1999: 6)

Lastly, the Klein Grid was easily interpretable for the researcher with statisti-cal tools, but it did not necessarily provide numbers that were meaningful to the test-taker.

Whatever their awareness of the Grid's flaws, Klein and two junior col-leagues tested it with data from the nearly 600 men and women who answered the *Forum* magazine call for participants. They conducted various two-way statistical analyses on their data, including a Friedman Two-Way Analysis of Variance, a Hotelling T-Square test, and a Canonical Correlation Analysis to investigate relationships among the different elements. Klein and his col-leagues found that:

> [t]he sexual orientations of the individuals in this study often changed remarkably over the period of their adult lives. All three of the self-identified groups [homosexual, heterosexual, and bisexual] became sig-nificantly more homosexually oriented over time … Many are potentially capable of travelling over a large segment of the sexual orientation continuum.
>
> (Klein et al. 1985: 45)

Klein found that his subjects' sexual orientation was not static, and that their learning about gay, lesbian, and sexual liberation movements probably played a role in his subjects' shift toward homosexuality. Klein argued that his Grid modeled better than the Kinsey Scale for the seven different variables he identified that made up a person's overall orientation: the sexual self (attraction, fantasy, and behavior), preferences (emotional, social, and lifestyle), and self-identification. Both the Kinsey Scale and the Klein Grid ended with assigning an individual a single number to describe that person's sexuality, but Klein thought that his number provided both the researcher and the subject more specificity and nuance. As Klein and his co-authors wrote, "two people with an overall average of '4' for the 'present' are very different if one has the configuration 4-4-4-4-4-4-4, while the other has 2-1-3-6-7-5-4" (ibid.: 47). In the end, however, the researcher and the participant were still left with a single number to capture an individual's sexual complexities.

Moreover, Klein and his associates used the *Forum* magazine results to criticize strongly the still-current three-part labeling system for sexual identity, arguing that those three labels did not adequately represent the diversity of sexualities that many people lived every day. Asexuality was also left off of the Klein Grid. He did not offer a solution to that conundrum, but rather encouraged his fellow scientists "to be more explicit in describing which aspects of sexuality and emotional/social preference are being considered as variables, and to use a multivariate design rather than a simple contrast of distinct groups" (Klein 1978: 48). Klein's scientific readers did not take his results very seriously, as for them self-selected readers of a pornographic magazine like *Forum* did not constitute an adequate sample population. But they were attracted to a method for analysing sex-related data that provided the opportunity to utilize ever-more sophisticated computer-based statistical techniques.

After the Klein Grid was published in the *Journal of Homosexuality* in 1985, sex researchers adopted it and adapted it for their own purposes. The Grid gained further scientific attention after a team of medical and psychiatric researchers affirmed its external validity in 1993 by testing it with subject populations in New York and California (see Weinrich et al. 1993). Other researchers have used the Klein Grid to ask a wide variety of people questions about their sexual pasts, presents and futures. It was easy to administer, fun, and simple to take, and the data can be easily manipulated in multiple ways statistically. Over time, researchers have modified the language of the questions and have often set aside the questions on political identity.[6]

It has also inspired other tests of sexual identities, most prominently the 45-question Multidimensional Scale of Sexuality and the Measure of Sexual Identity Exploration and Commitment (MoSIEC), a 25-item scale with four distinct factors common across all sexual identities: exploration, uncertainty, commitment, and synthesis integration. MoSIEC is the only sexual identity measure of 218 total measures in the third edition of *The Handbook of Sexuality–Related Measures* (2011), the standard guidebook for sex researchers using quantitative methods (see Worthington et al. 2008; Thompson and

Morgan 2008; Fischer et al. 2011). So the Klein Grid continues to be useful in and of itself for sex researchers looking for information specifically about issues of sexual identity and orientation, and as an inspiration for further tests. The absence of other professionally accepted sexual identity tests in the *Handbook*, however, speaks to the difficulty of creating a diagnostic tool that satisfies the short- and long-term interests of clinicians and their clients regarding frameworks for considering, deepening, and integrating an individual's sexual identity.

The Klein Grid has also been easily adaptable to the Internet, so that anyone—not just people interested in sex research projects or in attaining clinical assistance with their sexual identity—could take it. Websites in several different languages offer original or modified versions of the Klein Grid that instantly translate one's answers to the questions into a number with two decimal points. The most popular version of the Klein Grid is on a British website providing online and in-person youth and family services.[7] Klein Grid quizzes exist online in multiple languages including English, French, and Dutch. A Klein Grid quiz is easy to do online, and the results are anonymous and not stored anywhere after completion. However, it has not become popularly discussed online the way that the Kinsey Scale and made-up quizzes about the Kinsey Scale have been on websites about sexuality, spirituality, teen issues, and self-help (see Drucker 2012). Those websites that do have places for user response about the Klein Grid tend to treat it lukewarmly and with skepticism.

An anonymous writer rewrote the Klein Grid in their own words as a quiz on the free American dating website OKCupid around 2008. When a member of OKCupid finishes the 18-question quiz, which included some questions unrelated to the Klein Grid, the quiz calculates the user's answers into one of the seven Kinsey numbers. For example, if the quiz-taker gets a Kinsey 1, he or she gets a message with Kinsey's original language ("predominantly heterosexual, only incidentally homosexual") followed by a colloquial interpretation of that number: "Basically straight but with a slight turn to the road. Like most people you are fairly constant in your attentions but your eyes have occasionally strayed. Some members of your own gender are just so pretty." If the quiz-taker gets a Kinsey 4 ("predominantly homosexual, but more than incidentally heterosexual"), the message is: "You could be gayer. You're like a rubber eraser. You might look solid but you are a bit more flexible than average. You have a leaning towards members of the same sex but generally you love people across the board."[8] This fairly accurate rewrite, blending the two best-known sexual identity research measures, was aimed toward helping OKCupid users find better matched friendship, relationship, and sexual partners, not toward personal self-discovery, although some users may have found insight along the way.

In addition to that modified quiz, the Klein Grid is posted on a wide range of sex- and gender-oriented websites, there are also dozens of English-language blogs that posted the author's results of the Klein Grid. The majority post their quiz results without further comment. Across all the blog entries and the comments that follow them, most people who take a Klein Grid quiz are dissatisfied with it and with their results. For example, a user named Josh on Snark Forums wrote in 2008: "You know, this doesn't really tell you anything you couldn't already figure out. I mean, if you are currently having sex with the same gender or you would be ideally, chances are you're at least a little gay."[9] A handful of bloggers have thought through the implications of the Klein Grid, and most of them (along with their commenters) found it to be a moderately helpful tool in contemplating their sexual identity, but only as one among other methods of doing so.

For example, blog author Sonya Lynn wrote about the Klein Grid in 2009:

> It was meant to be an improvement on the mono-numerical Kinsey Scale, at which it does succeed admirably, but one could easily imagine a much more thorough modeling after another 31 years to ponder the interactions between sex, gender, and attraction. To someone like us, the Klein Grid seems as quaint as the Kinsey Scale was to Klein, but to a lot of people, the notion that your sexual orientation could a) be multi-faceted, and b) change over time, is still a revolutionary notion outside of bubbles like San Francisco, Seattle, New York City, etc. It's, if anything, not about labeling, but about challenging labels. It's a preliminary step in making people think of their sexual identity as more than just "gay" or "straight".[10]

Lynn noted that the Klein Grid, like the Kinsey Scale, could become more easily available to individuals outside of major, traditionally liberal cities thanks to the Internet. She also argued, as did other bloggers, that the specific numbering that a Klein Grid quiz yielded was less important than the fact that people were contemplating their sexual identity seriously. Blog reader "moi" replied to Lynn's original post, highlighting the flexibility and change-ability of the scale's results for an individual over time:

> The Klein Grid isn't intended to be one-stop shopping for orientation. In fact, the assumption is that each time you fill one out all three facets should look at least slightly different each time. The intention was to create a more fluid way of understanding an individual's mindset at any given moment without insisting that "feeling like a 7" has any long-term value ....[11]

The commenter Jim@HiTek on a 2006 post on the blog Goblinbox shared a similar opinion regarding the blog author's Klein Grid result of 2.38:

> Humans swish (like that word in this context) between things like that as the wind blows and their needs change or their situations with the humans around them change. You don't like labels ... remember? Well,

[More Than Incidentally Homosexual] is just a newer label that might fit now, won't fit in a month. (Probably.)[12]

Bloggers and Internet writers have also noticed how the Klein Grid counted same- or opposite-sex friendship as an aspect of sexual orientation and usually disagreed with it. The user "fervour" on the blog Straightdope took issue with how the Klein Grid took opposite-sex friendships into account:

I find the quiz results to be suspect. I scored 4.67. I'm gay. Because I have and have always thought that it's good to form strong emotional bonds to members of the opposite sex and because I have and have always felt that it's good to hang out in mixed company, I'm between "more than incidentally" and "incidentally" hetero? Playing with the questions: Switching to strong same-sex emotional bonds only yields a result of 5.1. Basically, the only way to be a pure homosexual is only to associate with homosexuals. I think that's simply wrong.[13]

This writer thought that mixed-sex friendship did not make him or her any less gay. Klein did not describe the reasons why he included friendship in his sexual orientation calculation, although it is notable that he took friendship into account in the first place (which the Kinsey Scale does not do all). Perhaps it reveals Klein's view that gay, lesbian, and bisexual communities were strongest when they were separate, a view that present-day Internet writers on sexual orientation generally do not share (see Weinrich 1993: 166).

Other Internet users have problems with the scale as a whole, and some had serious criticisms of the Kinsey Scale and the Klein Grid's numerical labeling system, but only when outsiders applied it to individuals. For example, the blogger Androgynous Fox wrote a post called "Problems with the Kinsey Scale" in which he also addressed the Klein Grid:

There is harm in mislabeling. To identify as something and not be able to be taken seriously for one's identity (as still happens with bisexuality), or to be labeled by someone else with no chance at recourse is definitely a painful experience. It's quite nice to be able to specify what you are as best as you are able, even if that means providing one number or letter out of a scale (or, as proposed with the Klein Sexual Orientation Grid, 21 numbers that, alas, still leave some out) or nothing at all.[14]

Writers like Androgynous Fox appreciated the Kinsey Scale or Klein Grid, but only when they had control over the instruments and results themselves.

One aim of the original Kinsey Scale was to unlink sexuality from identity, as Kinsey saw much harm done to any individual labeled "homosexual" by outsiders, whether that person had a single homosexual experience or thousands. By using a number on the 0–6 scale to stand as an identity marker, Kinsey aimed to remove the stigma and punishment from labeling and providing an alternative way of knowing one's sexual self (see Kinsey et al. 1948: 610–66). Klein aimed his grid at individuals seeking to understand better their sexualities with the assistance of professionals. Neither researchers' reconceptualization of sexual labeling, however, has altered the present-day dominant cultural understanding of sexual identity as one of three options: straight, gay or bisexual. Although those three terms are highly debatable and leave a lot of people out, and "we cannot make our sexual concepts do all the descriptive and analytical work we need them to do," they remain widely used and popular, and their use by the general public seems unlikely to change (Halperin 2009: 455).

Numerical measurements of sexuality can be useful for health professionals, but they alone cannot change the language and concepts used to describe it. It is clear that although the Klein Grid is easily adaptable to an online quiz environment, it is ultimately a researcher's tool. The Kinsey Scale, although Kinsey originally conceived it as a research tool, now has a vibrant online life as an individual discovery tool.

Measurements for sexual behavior, identity, fantasy, and arousal are often unclear about what they are measuring, and to what end. As Amanda Udis-Kessler wrote about sex surveys in particular: "Surveys, questionnaires and other self-reporting methods tell us only about someone's self-perception, not necessarily about her behavior, motivations or unconscious influences" (Udis-Kessler 1999: 53). Of course, Klein himself realized that, as he wrote earlier in *Bisexual Option:* "The most important thing to note is that the bisexual, the heterosexual [and] the homosexual each lives in a state of motion, and within that state anything in the psychosexual spectrum is possible" (Klein 1978: 22).

What "sexuality" even is as a whole remains fuzzy and undefined: fluid, movable, debatable, perhaps ultimately unquantifiable, but that will not stop researchers from inventing and testing new tools to see which one might stick in the academic and possibly public imagination. Orientation, pleasure, body image, identity, fantasy and behavior are all parts of an individual's sexual life over time, and it is difficult to find a single tool that captures them all. As a satisfactory definition of "sexuality" remains elusive, tools to measure it will remain insufficient and incomplete.

Notes

1 American Institute of Bisexuality, "About Dr. Fritz Klein, MD", at http://www.americaninstituteofbisexuality.org/fritz-klein/, accessed December 18, 2016.
2 American Institute of Bisexuality, "About Dr. Fritz Klein, MD" at http://www.americaninstituteofbisexuality.org/fritz-klein/, accessed December 18, 2016.
3 J. R. Sylla, personal communication to the author, April 10, 2014 (in author's possession).
4 American Institute of Bisexuality, "The Klein Sexual Orientation Grid" at http://

www.americaninstituteofbisexuality.org/thekleingrid/, accessed 10 April 10, 2014, emphasis in original.

5 See for example *Love Your Enemy? The Debate between Heterosexual Feminism and Political Lesbianism* (1981); Cruikshank (1992); Abelove (1993).

6 J. R. Sylla, personal communication to the author, November 29, 2013 (in author's possession).

7 Young Southampton, "Klein Sexual Orientation Grid – Online Quiz'" at http://www.youngsouthampton.org/children-and-young-people/advice/relationships/sexuality/klein-sexual-orientation-grid-quiz.aspx, accessed April 10, 2014.

8 OKCupid, "The Klein Sexuality Test" at http://www.okcupid.com/tests/the-klein-sexuality-test, accessed December 18, 2016; The Kinsey Institute, "Kinsey's Heterosexual-Homosexual Rating Scale" https://kinseyinstitute.org/research/publications/kinsey-scale.php, accessed December 18, 2016.

9 Josh, "Klein Sexual Orientation Grid" at http://snarkforums.com/index.php?topic=6821.25, accessed April 10, 2014.

10 Sonya Lynn, "Well File This under 'DUH!'" http://www.sonyalynn.com, accessed March 10, 2013, no longer available.

11 ibid.

12 Jim@HiTek, comment on Goblinbox, "More Than Incidentally!" at http://www.goblinbox.com/archives/1321, accessed April 10, 2014.

13 Fervour, "Ever Fallen Head over Heels for Someone of the Wrong Sex?" at http://boards.straightdope.com/sdmb/showthread.php?t=534625, accessed April 10, 2014.

14 Androgynous Fox, "Problems with the Kinsey Scale" at http://androgynousfox.tumblr.com, accessed April 10, 2014.

References

Abelove, H., Barale, M. A., and Halperin, D. M. (eds.) 1993. *The Lesbian and Gay Studies Reader*. New York: Routledge.

Cruikshank, M. 1992. *The Gay and Lesbian Liberation Movement*. New York: Routledge.

Davis, K. B. 1929. *Factors in the Sex Life of Twenty-Two Hundred Women*. New York: Harper & Bros.

Dickinson, R. L. and Beam, L. E. 1934. *The Single Woman: A Medical Study in Sex Education*. Baltimore: Williams & Wilkins Company.

Dickinson, R. L. and Beam, L. E. 1932. *A Thousand Marriages: A Medical Study of Sex Adjustment*. New York: Century Company.

Drucker, D. J. 2012. Marking Sexuality from 0–6: The Kinsey Scale in Online Culture. *Sexuality and Culture* 16:3: 241–62.

Drucker, D. J. 2010. Male Sexuality and Alfred Kinsey's 0–6 Scale: Toward "A Sound Understanding of the Realities of Sex". *Journal of Homosexuality* 57:9: 1105–23.

Fischer, T. D., Davis, C. M., Yarber, W. L., and Davis, S. L. 2011. *Handbook of Sexuality-Related Measures*. 3rd edn, New York: Routledge.

Gooß, U. 2008. Concepts of Bisexuality. *Journal of Bisexuality* 8:1–2: 9–23.

Halperin, D. M. 2009. Thirteen Ways of Looking at a Bisexual. *Journal of Bisexuality* 93:4: 451–5.

Hamilton, G. V. 1929. *A Research in Marriage*. New York: Alfred and Charles Boni.

Kinsey, A. C., Pomeroy, W. B., Martin, C. E., and Gebhard, P. H. 1953. *Sexual Behavior in the Human Female*. Philadelphia: W. B. Saunders.

Kinsey, A. C., Pomeroy, W. B., and Martin, C. E. 1948. *Sexual Behavior in the Human Male*. Philadelphia: W. B. Saunders.

Klein, F., Sepekoff, B., and Wolff, T. J. 1985. Sexual Orientation: A Multi-Variable Dynamic Process. *Journal of Homosexuality* 11:*1–2*: 35–49.

Klein, F. 1993. *The Bisexual Option*. 2nd edn New York: Haworth Press.

Klein, F. 1978. *The Bisexual Option: A Concept of One Hundred Percent Intimacy*. New York: Arbor House.

Likert, R. 1932–33. A Technique for the Measurement of Attitudes. *Archives of Psychology* 22: 5–55.

Love Your Enemy? The Debate between Heterosexual Feminism and Political Lesbianism. London: Onlywomen Press 1981.

Sanders, S. A., Hill, B. J., Yarber, W. L., Graham, C. A., Crosby R. A., and Milhausen, R. R. 2010. Misclassification Bias: Diversity in Conceptualisations about Having "Had Sex". *Sexual Health* 7:*1*: 31–4.

Sanders, S. A. and Reinisch, J. M. 1999. Would You Say You "Had Sex" If…? *JAMA* 281:*3*: 275–7.

Storr, M. 1999. Editor's Introduction. In: *Bisexuality: A Critical Reader*, edited by M. Storr, 1–12, London: Routledge.

Thompson, E. M. and Morgan, E. M. 2008. "Mostly Straight" Young Women: Variations in Sexual Behavior and Identity Development. *Development Psychology* 44:*1*: 15–21.

Udis-Kessler, A. 1999. Notes on the Kinsey Scale and Other Measures of Sexuality. In: *Bisexuality: A Critical Reader*, edited by M. Storr, 49–56, London: Routledge.

Weinrich, J. D., Snyder, P. J., Pillard, R. C., Grant, I., Jacobson, D. L., Robinson, S. R., and McCutchan, J. A. 1993. A Factor Analysis of the Klein Sexual Orientation Grid in Two Disparate Samples. *Archives of Sexual Behavior* 22:*2*: 157–68.

Worthington, R. L., Navarro, R. L., Savoy, H. B., and Hampton, D. 2008. Development, Reliability, and Validity of the Measure of Sexual Identity Exploration and Commitment (MoSIEC). *Developmental Psychology* 44:*1*: 22–33.

Internet Sources:

American Institute of Bisexuality, About Dr. Fritz Klein, MD at http://www.american instituteofbisexuality.org/fritz-klein/, accessed December 18, 2016.

American Institute of Bisexuality, The Klein Sexual Orientation Grid at http://www. americaninstituteofbisexuality.org/thekleingrid/, accessed December 18, 2016, emphasis in original.

Androgynous Fox, Problems with the Kinsey Scale at http://androgynousfox.tumblr. com, accessed April 10, 2014.

Fervour, Ever Fallen Head over Heels for Someone of the Wrong Sex? at http://bo ards.straightdope.com/sdmb/showthread.php?t=534625, accessed April 10, 2014.

Jim@HiTek, comment on Goblinbox, More Than Incidentally! at http://www.goblin box.com/archives/1321, accessed April 10, 2014.

Josh, Klein Sexual Orientation Grid at http://snarkforums.com/index.php?topic=68 21.25, accessed April 10, 2014.

Lynn, Sonya Well File This under "DUH!" at http://www.sonyalynn.com, accessed March 10, 2013, no longer available.

OKCupid, The Klein Sexuality Test at http://www.okcupid.com/tests/the-klein-sexu ality-test, accessed December 18, 2016.

The Kinsey Institute, Kinsey's Heterosexual-Homosexual Rating Scale https://kinsey institute.org/research/publications/kinsey-scale.php, accessed December 18, 2016.

Young Southampton, Klein Sexual Orientation Grid—Online Quiz at http://www. youngsouthampton.org/children-and-young-people/advice/relationships/sexuality/ klein-sexual-orientation-grid-quiz.aspx, accessed April 10, 2014.

13 The Desert and the Dendrograph: Place, Community, and Ecological Instrumentation

Emily K. Brock[1]

Inside a small stone workshop on a rocky hillside outside of Tucson, Arizona, surrounded by an exquisite yet forbidding landscape of desert plants, Godfrey Sykes created scientific instruments. An employee of the Carnegie Desert Laboratory, his workshop sat adjacent to the main laboratory building. Both buildings baked in the dusty heat of an Arizona summer. The Desert Lab ecologists sometimes found their arid, remote field sites less hostile to humans than the oppressive heat inside the complex of structures where they lived and worked.

It was 1908, and the Desert Lab had only recently been wired with rudimentary electricity; Sykes still drilled metal by hand and tightened fittings under natural light. The electricity, the plumbing, the water well, the road to the Lab, and indeed the buildings themselves, had all been established by Godfrey Sykes. Inside his small shop, Sykes went about his tasks, fabricating anything that the Carnegie Desert Lab's ecologists could need to conduct research. Sykes would marvel at how

> [m]anifold their requirements were for the proper prosecution of their absorbing science ... A colleague would lean over one's desk or work-table, and ask for information or advice as to where to get a flat-iron, a piece of raw-hide, an empty barrel, or a looking glass, which were urgently needed for the completion of some experimental work upon which he was engaged.
>
> (Sykes 1945: 280)

Sykes tried to accommodate all requests as quickly and cheaply as possible, with the benediction of the president of the Carnegie Institution of Washington, who had stated his "wish and intention that the research workers should have every facility and appliance that they needed for their investigation provided for them" (ibid.). To meet any possible need, the shop was stocked with "a full complement of carpenter's and cabinetmaker's tools, lathe, drill press, band saw, planer, forge, grinding wheels, etc.,"[2] along with spare parts for repairing vehicles and equipment. Sykes was also a vital asset away from the lab complex, repairing broken-down automobiles during field expeditions in remote deserts, and building small boats for geographical and botanical surveys of the Salton Sea and the Colorado River Delta.

The Desert Lab's harsh climate inspired innovation. It was the birthplace of a number of inventive field instruments, tools which facilitated its researchers' fusing of plant physiology and field ecology. Cross-disciplinary exchange was brought about by scientists' forced intellectual intimacy in the isolated and primitive lab, and by the lab's proximity to the desert environment itself. Their innovative theoretical approaches to the field were manifested in new forms of instrumentation as well as in the research papers they generated. The ability to collaborate across disciplinary boundaries to fabricate and refine new instruments led them to new research directions. Collaborative work challenges notions of authorship, scientific or otherwise. Nonscientist laboratory workers have been characterized as "invisible technicians," whose work has been left out of both scientific and historical accounts of academic research. Godfrey Sykes, however, was a noninvisible technician, one whose work was so important to the laboratory's goals that his imprint can easily be found (see Shapin 1989).[3] The instruments he helped create not only facilitated particular approaches to ecological research, but also embodied new theoretical directions for the field. This piece examines the collaborative technical community at the Desert Lab, with particular focus on how one of those tools, the dendrograph, fostered a reliance on accurate measurement that demonstrated the scientific standing of ecology.

Building Community for Collaboration

Godfrey Sykes became the central node of collaboration in the Desert Lab, involved with almost every project and consulted on almost every problem. Sykes was one of the first employees of the Carnegie Desert Lab, and arguably its most important employee after the director, Daniel Trembly MacDougal. However, Sykes was not a trained ecologist at all, but rather a man with vast abilities to make, fix, and design, coupled with untiring curiosity and a love of exploration. The Desert Laboratory had been founded in 1903, with MacDougal appointed as its first real director two years later, and Sykes had been the first permanent employee MacDougal had hired (see Colville and MacDougal 1903; Sonnichsen 1987; Bowers 1990; MacIntosh 1983; Kingsland 1991; Bowers 1988; Kingsland 2010).

McDougal found in Godfrey Sykes a man uniquely qualified to assist in the creation of the Desert Lab. Sykes, trained as an engineer in his native England, had immigrated to the United States where he was first a New Jersey farmhand, then a cowboy in the central plains. After an interval in Japan in the employ of an American hydraulic engineering firm, he had joined his brother Stanley on the outskirts of Flagstaff, in the northern mountains of Arizona. By the late 1890s, the brothers were devoting much of their energy to a side-business in mechanical engineering, which he advertised as

"Makers and Menders of Anything." Sykes's engineering skills were put to use in various local machining and engineering projects, from retrofitting hotels with interior plumbing to rigging gambling machines. The Sykes shop also hosted Flagstaff's local community of scientists and naturalists, who "organized themselves into a sort of informal club, under the name 'The Busy Bees,'" for "social intercourse and the exchange of current scientific gossip" (Sykes 1945: 237).[4]

It was in Flagstaff that Sykes had first met MacDougal, during a federally-funded botanizing expedition in the early 1890s. MacDougal then hired Sykes to accompany him on extensive scientific surveys of the Colorado River Delta during the winters of 1904 and 1905, trips which cemented the men's friendship (see Colville and MacDougal 1903: 38). MacDougal was in many ways the opposite of Sykes. With a doctorate from Purdue University and postdoctoral study in Europe, he had first become interested in desert plants as an associate director of botany at the New York Botanical Garden.

While the New York Botanical Garden provided a comfortable niche for a research career, MacDougal chose to strike out for new horizons by signing on for the responsibility of establishing this new desert enterprise. MacDougal knew Sykes would be an invaluable asset to nurture the Desert Lab into a successful research institution, as establishing an advanced research facility in that remote area would require not just prominent scientists but also practically-minded problem solvers. The shop at the Desert Lab reconstituted the mood of Sykes's Busy Bees, but now backed with the significant financial support of the Carnegie Institution and energized by the common aims of the scientists gathered there (see Sykes 1945: 263). Sykes's official role was the facilities and maintenance manager of the Desert Lab, but his deep knowledge of the desert's Southwest geography and climate was also a valued asset. Along with Sykes, MacDougal recruited researchers and technicians, either as full time Carnegie employees or on research leave from academic positions. His aim was to populate the lab with people who had experience in ecological field work and in plant physiology.

MacDougal justified Sykes's hire, and the comfortable salary the technician was afforded, by pointing to the need for self-sufficiency to conduct scientific research in remote and rugged Tucson (see Sykes 1945).[5] But while that might have been the proximate cause for Sykes's hire, MacDougal also knew that having him there could nurture an inventive and original atmosphere for research at the Lab. Having access to a man like him not only meant that the researchers could have things they could not otherwise find or afford, but that they could customize them to their research needs and to the harsh desert climate. Also, the shop reduced costs associated with sending out for the fabrication and repair of machinery and equipment. But more importantly, keeping the shop as part of the Lab complex meant the scientists could go through a process of collaboration with Sykes, fine-tuning instruments to fit their needs exactly. Delegating fabrication duties to faraway contractors, or ordering pre-made instruments from established East Coast instrument making firms, also would not have allowed for spontaneous, across-the-table conversations in Sykes's workshop.

In time, the instruments Sykes created at the Desert Lab included highly complex tools for ecological fieldwork. One of the most complex and creative was the dendrograph, which measured trees' physiological responses to changes in their environment. The physical proximity and social interaction at the Desert Lab allowed its researchers to collaborate with Sykes and each other. Desert Lab researchers' hybrid tools, modifications, and entirely new inventions made it possible to measure phenomena hitherto only studied in laboratories, pushing the nascent field of ecology in new directions. The new instruments they made, tailored to the specific questions the researchers had, pushed the nascent field of ecology in new directions.

This laboratory was not the austere, controlled environment laboratories were in other localities. The conventional barriers between indoors and outdoors did not hold in early-1900s Tucson. The occasional corollary assumption by historians, that indoor laboratories are more comfortable than outdoor field sites, also did not hold in this time and place.[6] Many staff and scientists elected to live in wood-framed, canvas-sided tents on the Laboratory grounds, rather than suffer the heat of conventional homes. Laboratory buildings are always tools for science, and the hostility of the Desert Lab's environs demanded more innovation from these tools than most laboratories did. The Lab's main building, built of the same roughly-hewn volcanic rock that composed the mountains, had large windows, roof vents, and wide eaves in an attempt to cope with the desert heat. Sykes also designed a constant-climate plant room dug into the bedrock, which promised less diurnal variation in temperature. Despite these innovations, the fact remained that for much of the time the scientists spent working indoors, they would be more uncomfortable than if they were working in the fresh air. Accordingly, their research turned outwards, towards the desert.

The Desert as Location and Subject of Study

The Desert Lab was one element in a wide-ranging project to assist many new, promising directions in American science. The Carnegie Institution focused on helping scientists simply generate basic research, without any role for education or utility. They funded scientific enterprises as diverse as Cold Spring Harbor's genetics and evolution studies in New York, astrophysical study at California's Mt. Wilson Observatory, and the creation of an entirely metal-free sailing ship to plot variations in the terrestrial magnetism of the Earth. Recognizing that ecology was poised to revolutionize the study of natural biological systems, they funded ecological work of many kinds. They underwrote Luther Burbank's plant breeding research on the Monterey Peninsula of California, the work of Frederic Clements's Alpine Lap near Pikes Peak, Colorado, and a marine biological laboratory on the remote Loggerhead

Key in Florida's Dry Tortugas.[7] The Carnegie Institution focused on guiding scientific enterprise and setting research agendas, as well as furnishing scientists with generous operating budgets.

As director of the Desert Lab, MacDougal was charged not just with administering a research facility, but with setting goals to propel plant ecology in innovative directions. The Desert Lab was the ideal place to foster new directions in ecology, to jolt the field out of its ruts. With the budget and focus to support ecological innovation and few of the distractions found at a university, ecologists could try risky and uncertain ideas with fewer worries. At its founding, the Desert Lab was heralded as an institution that would allow, through its focus on field studies of plant physiology, the discipline of ecology to break free of the habits that were holding it back.

The Desert Lab would also generate large scale plant geographical projects of the deserts, in-depth regional studies of rainfall and climate, the creation of permanent scientific plots of desert plants, numerous exploratory surveys and mapping expeditions of the Colorado River Delta, and a study of the newly-formed Salton Sea as a giant ecological experiment (see McGinnies 1981; Sykes 1945; Bowers 1990). One of the most important of the Desert Lab's research directions was the one in which MacDougal himself was most involved, the development of methods for understanding the day-to-day physiological response of plants to the environment in which they live.

Darwin's vision of a natural world eternally at war, where each living thing struggled against both its competitors and against the harsh limits of environmental conditions, was an essential element to his theory of evolution by natural selection. Understanding ecological relationships would allow scientists to gain insight into the long-term processes of evolution by natural selection. In the first years of the 20th century, the numbers of biologists calling themselves ecologists was on the rise. The influx of new researchers caused reassessment of methods and theory. In the 1890s, American plant ecology had developed as a refinement and refocusing of the diffuse pursuits of natural history and plant geography. Many early American plant ecologists focused heavily on the study of plant distribution, using techniques of survey and census to create textured but ultimately static descriptions. Completing such studies required little more than basic tools of measurement coupled with lavish amounts of labor.

Research in the founding decades of the discipline often consisted of exhaustive ecological portraits of plant associations mapped onto localities' soils and climate (see Tobey 1981; Kohler 2002). Plant surveys could not show dynamic change, describing rather species distribution, morphology, and corresponding climate and topography. From this information, researchers might extrapolate conclusions about ecological dynamics, but the dynamics were not examined directly. Such studies were not necessarily simplistic, with ecologists such as Henry A. Gleason producing groundbreaking results without instrumentation or experimentation (see Nicolson 1990; Nicolson and McIntosh 2002). While surveys contributed much to American knowledge of plant diversity and distribution, they did little towards fostering understanding of the dynamic Darwinian struggle for existence.[8]

The Desert Laboratory was founded to look forward to the future of ecology, not to accommodate the comfortable standard practices of the field. MacDougal, with help from the administration, molded the Lab to fit his and his allies' goals in ecology, and in turn to support and inspire other ecologists with its devotion to the cutting edge. In its early years, the Desert Lab did more than just provide a place for research; it also did much to set new agendas for the field of plant ecology as a whole. Investigation of the interactions between species and their surroundings, especially as regards adaptation by the species to the realities of environmental conditions, was more difficult to undertake.

Observers heralded the Desert Lab as indispensable to the development of such a physiological ecology, however, by prompting both workable methodology and instrumentation to conduct such research.[9] The synthesis of "environmental physics and adaptational physiology," one scientist wrote approvingly upon the Lab's opening, was the only way for ecology to continue making strides forward as a science that was to focus on "this indispensable dual study" (Ganong 1904: 494).[10] Indeed, he went on, such research would be "the very soul of ecology" (ibid.: 493).

One long-term associate of the Desert Lab was plant physiologist Burton Livingston, reflecting MacDougal's resolve to foster a physiologic approach to plant ecology. Although plant physiology traditionally involved little field work, Livingston would maintain his relationship with the Desert Lab for more than a decade, and would prove one of its most creative and prolific researchers. Plant physiologists were tethered to the greenhouse and laboratory, reliant on a battery of complex instruments to push forward research on the conditions of plant growth. Observations of climate conditions could be considered alongside results of laboratory studies of physiology, but no satisfying conclusions about real plants in real deserts would be reached with that method. Livingston recognized that advancing physiological ecology depended on scientists finding ways to carry equipment outdoors and conduct physiological research in the field (see Nicolson 1990; Livingston and Lawrence 1948; Went 1957). Although the lab building had been designed to maximize natural cooling, the dry desert heat still invaded the Desert Lab's buildings to the extent that it could affect experiments done within the walls. On the other hand, the harsh deserts of southern Arizona were particularly good for field study of physiological ecology. Because the local climate and soil were so hostile, plants were usually spread out sparsely on the landscape. Ecologists found isolating physiological factors difficult in densely packed landscapes such as forests and marshes, where plants competed for access to sun, water, space, and nutrients. In the desert, the physiological effects of the abiotic environment could be easily isolated, since the effects of biological competition were minimal. Further, physiological effects of extreme climate challenges could be easier to study than subtle effects. The desert environment thus presented scientists with both a challenge and an opportunity, inspiring them to invent new, custom-designed instruments to study physiological ecology (see Went 1957).

The Dendrograph as Collaborative Innovation

The Desert Lab took preexisting laboratory instruments of plant physiology, and refined and reimagined those instruments to address ecological questions. These new instruments allowed the accumulation of innovative research at the Lab, but they also shifted the discipline of plant ecology overall, towards a science that intersected more with the theoretical aims of evolutionary biology and less with the descriptive aims of natural history. Sykes's workshop created a number of one-off instruments for the Desert Lab, designed to meet specific researchers' proclivities or their needs for a specific experiment. But several instruments designed by affiliates of the Desert Lab found a larger audience based on the research papers in which they featured. The scientists arranged for manufacture of these instruments, including the dendrograph, which they sold to the wider community of ecologists. By promoting their distribution, the Desert Lab researchers promoted specific approaches to theoretical questions in ecology as well as a rigorous and technological-mediated approach to ecological fieldwork. Sykes remained the central node of collaboration but as the research and tools became more complex the collaborators' conversations often moved beyond that workshop table. Sykes, never a mere "invisible technician," maintained an increasingly visible role as the lab's expedition leader and geographer (see McGuire 2003; Sykes 1937).

In December of 1908, the Desert Lab staff had an unexpected, but welcome, piece of news from California. Luther Burbank had grown increasingly erratic and uncommunicative at his experimental horticultural operation near Carmel-by-the-Sea, on the Monterey Peninsula. As a result, Desert Lab staff had been sent to monitor and manage Burbank's work during the summers of 1907 and 1908. Perhaps impressed by those visitors, or wishing their nascent town to be known as an intellectual center, The Carmel Development Company offered to build for the Carnegie Institution an ecological laboratory on the model of the Desert Lab. President Woodward accepted this generous donation, but rather than earmark the windfall for the out-of-favor Burbank, he appointed MacDougal to manage it as an extension of the Desert Lab.

By the summer of 1908, the Coastal Laboratory was up and running. MacDougal and his staff began to establish new research centered on the ecology of the Pacific shore, especially in the forested hills above the town. While the Coastal Lab had the undeniable benefit of being far from the desert heat, it was also unfortunately far from the scientific community at the main lab in Tucson. While in Carmel, MacDougal corresponded with Livingston or Sykes twice a week or more, monitoring research activities in the desert as well as consulting with them on his own scientific interests. His letters to Sykes often included ruminations on specific ecological questions, and Sykes's letters back to him often included sketches of instrument designs which could probe those same questions. The arrangement increased the research capacity of the Desert Lab, but the Tucson facility remained central (see Craig 2005; Bowers 1990; Kingsland 2010).

It was while in Carmel that MacDougal began to develop an interest in how trees' daily physiological cycles responded to environmental phenomena. Studying transpiration in the field is a way to directly record the interaction of plants with their environment. Plants draw water up through the roots and expel it through the leaves in the process of transpiration. The process is a necessary part of photosynthesis, and is also the way the plant responds to heat, aridity and other environmental pressures. The expelling of water vapor occurs through the leaves' stomata, which are open in daylight and closed in the dark. While leaf transpiration rates can be easily monitored on small plants in the lab, it was not feasible to make such measurements on trees in the field. Moreover, the quantity of water vapor expelled by trees induces fluctuations in trunk girth as the volume of water flowing through the plant changes. Direct measurement of transpiration meant measurement of the interaction between the tree and the environment in which it was situated. Measurement of change, rather than measurement of successive static states, could contribute importantly to building an understanding of physiological ecology.

MacDougal conceived of a tool which could measure precisely the small daily fluctuations in the girth of tree trunks in response to environmental stimuli. This tool could be used to study the diurnal fluctuations in the tree, seasonal variations in response to changes in light and precipitation, and long-term developments in tree growth. The crowns of trees, governed by upper branches that swayed, moved, and broke off, varied greatly through the seasons and from year to year. The trunk, however, had a steadier rate of change. It grew incrementally wider each year, and hence served as a quick descriptor of the size and age of the tree.

The dendrograph worked by providing continual information about the girth of the tree as it reacted to environmental conditions. It was a hoop-shaped metal apparatus that adjusted to fit closely around the circumference of a tree trunk. As the trunk pressed against the apparatus, a measuring device indicated the minute, yet continual, changes in volume to the trunk of a tree during the daily cycle of transpiration. The most elaborate versions transcribed data onto a roll of paper, providing a continual record of the daily cycle of transpiration and the long-term climatic and seasonal changes in that cycle. The recording device meant data could be taken without constant attention by scientists. This solved one of the major issues with ecological fieldwork, the impossibility of recreating the workflow of a well-staffed laboratory. The visual production of the time chart was significant. The expanding and contracting of the trees could not be seen by the naked eye, but could be easily seen on the rolls of paper.[11]

The dendrograph had the potential to benefit MacDougal's own reputation and boost the Desert Lab's profile through promoting a physiological approach to ecology. The dendrograph MacDougal built in 1918 was unwieldy and inconstant, was still far from perfect, and had yielded few satisfying results. Nonetheless MacDougal had high hopes for it. He had built this first one with input from Livingston only, while Sykes had been away from his workshop on war-related leave in England.

In mid-June of 1919, in the address he read as the President of the Pacific Division of the American Academy for the Advancement of Science, MacDougal described the dendrograph he had made the previous summer and discussed its potential for generating new forms of data about tree physiology. That address, subsequently published in the Academy's journal *Science*, thrust the invention into the scientific limelight, but offered only a description of the instrument, and presented no results (see MacDougal 1919). From the results he had gotten in that first year, he got a sense of how the research could take place and pinpointed problems in the dendrograph design. In 1919, Sykes, returned from England, was again able to collaborate with MacDougal. He tested dendrographs, modified and replaced components deemed too delicate for weather conditions, found reliable sources for parts, and field tested the prototypes of new versions. The collaborative process continued until the two had arrived at a precise, rugged, reliable version of the device.[12]

The research generated through the use of a scientific instrument is more widely accepted within a scientific community if the instrument itself is well-known and widely used. MacDougal could have simply continued to publish papers based on his own monitoring of his set of dendrographs near Carmel. However, the dendrograph was valuable not only for its utility as a measuring device, but as a representation of the potential of physiological ecology. By distributing copies of the instrument widely, MacDougal would also gain comparative knowledge about different ecological settings.

By 1920, Sykes and MacDougal had a satisfactory mechanism for the dendrograph, and began distributing them to other Carnegie-funded ecologists. Through doing this, he got some sense of the physiological response of other tree and cactus species in various ecological settings in the West. As well, by learning what difficulties those scientists had in setting up and using the instruments, they could determine any further mechanical improvements needed. MacDougal and Sykes used this feedback to modify the original dendrograph design in subsequent rounds of collaboration. The second-generation dendrograph design was both easier to use and more adaptable to local variations in species and environments.[13]

In December of 1920, when he was fairly confident they had found the right design, MacDougal applied for and soon received a patent for the dendrograph. He made six more dendrographs between 1920 and 1924 for Carnegie-affiliated scientists in Canada, the US, and South Africa, and gathered feedback on their success.[14] In 1924, with an eye towards a larger distribution, MacDougal transferred the patent to the Carnegie Institution, which would allow easier sales and fabrication.[15] MacDougal supported the patent transfer. A wider use of the dendrograph could be beneficial to both the field of ecology and MacDougal's standing within it. As the Institution maintained an agenda of charitable scientific encouragement, it would set a nonprofit price for the instrument and help with sales and distribution.[16]

The Carnegie Institution had patented a number of instruments, of widely varying demand, for their scientists in various disciplines. The dendrographs were to be sold to scientists at cost, in line with the Carnegie's overall philanthropic goals of furthering basic scientific research. The sales orders were handled through the Washington office, which then paid the bills for labor and expenses (see Bowers 1992).

Both MacDougal and Sykes disliked the repetitive busywork of assembly, but MacDougal was loath to allow the dendrograph's manufacture out of his control. In 1919, the Carnegie Institution had approved, at MacDougal's request, a budget item of $2400 for the Desert Lab to hire a full-time "instrument maker and foreman" to construct the dendrograph and other scientific instruments for the lab's scientists. Unfortunately, skilled laborers were still rare in the Southwest, and no competent instrument maker could be found in Tucson.[17] Instead, the Desert Lab staff simply completed the work themselves. While MacDougal still directed the process, Sykes was now busy with both his own geographical research and the collaborative tinkering and engineering with the others in the lab (see Craig 2005; MacDougal 1914; Sykes 1937). In 1925, MacDougal still felt uncomfortable giving up the fabrication, writing that:

> I am not yet ready to advise that we license any manufacturer to take up the making of these instruments ... I am hoping that in another year or two we shall be able to make arrangements with some reputable manufacturer who could be trusted to attend to all of these details in making instruments for other people as well as for ourselves. At present the assembly of each instrument takes an appreciable part of my time.[18]

Godfrey Sykes had been integral in the design and creation of many of the lab's instruments, but was generally not involved in the repetitive busywork of assembly. MacDougal contracted a jeweler based in nearby Pacific Grove, whose main business was watch repair, to fabricate dendrographs for him on demand.[19] Sykes still machined most of the dendrograph's special parts in the shop in Tucson before shipping the parts to MacDougal in Carmel. The jeweler put them together at the direction of MacDougal, and was paid by the Washington office. Despite MacDougal's professed desire to have the instruments manufactured elsewhere, this would remain the standard procedure for their fabrication.[20]

MacDougal continued to expand his dendrographic studies in both the forests around Carmel and the Arizona deserts. The long-term studies, the first of which he had begun in 1918, now extended far enough to provide data not just about the seasonal and climatic effects on transpiration, but also about the incremental annual growth of the tree. He conducted trials involving multiple dendrographs at different heights on a single tree, and expanded the range of trees which the dendrographs had attached to in the areas around both the Coastal Lab and the Desert Lab. He also attached dendrographs to saguaro and barrel cactuses, which yielded striking data sets quite unlike those of woody trees. A confirmed experimentalist, he would eventually use dendrographs in a series of experimental bark removals to see the effect on transpiration, as well as in a study of the physiology of fire-damaged trees (see MacDougal 1930; 1936; 1946). His main constraint was the number of dendrographs he had on hand, which was in turn constrained by the time-consuming process of fabrication. As he plotted out his research plans in 1927, he mused that "this and other extensions of my experiments await the construction of additional dendrographs."[21]

Sykes continued to be involved with the creation of new dendrographs, both for the Desert Lab and for client orders. MacDougal relied on Sykes to create the clamps, rods, and bolts which made up the elements of the dendrograph, although the jeweler in Pacific Grove was still tasked with assembly. Sykes continued to track of the success or failures of the dendrographs, and modified existing parts and materials as needed.[22] For example, when a staff researcher noted in 1926 that the standard dendrograph was not satisfactory on the fast-growing cottonwood, Sykes was the one who determined the proper modifications to keep the instrument functioning.[23]

MacDougal's own fascination with the dendrograph's potential, and his finicky and specific design, hampered any attempts to streamline production. Although he complained of the time he spent with dendrograph sales, the design of the dendrograph was not standardized enough for him to delegate the job. On receiving an order, MacDougal often replied with inquiries about the species under study, the environment in which the dendrograph was to be used, and the scientific questions the buyer hoped to have answered. He needed this information to ensure they sent the right size, furnished with fittings appropriate for the trunk and bark on the trees under study, and in some cases supplied it with parts appropriate for harsh conditions.

His attention to these details was also due to his interest in physiological ecology, and his desire to see that field grow. Foresters and agricultural managers often had more money to spend on research equipment than did ecologists used to working with nothing more than a notebook and hand tools. MacDougal had some interest in the results of dendrographic research in agriculture and forestry, and understood in particular the promise of an ecologically-informed forestry. He also proved eager to send instruments to scientists overseas, often going to great lengths and incurring extra costs for the Institution.[24] This wide use would boost the dendrograph's fame, but could also provide MacDougal with new data on comparative physiological ecology.

Conclusion

Maintaining a bespoke fabrication process in Carmel and Tucson, rather than developing a universally-adjustable model manufactured offsite, doomed the instrument's chances to outlive its makers. Overall, around 50 to 70 dendrographs were created, ordered for research in basic ecology as well as for orchard and forest management. Although this is not an insubstantial number for a specialized instrument, there were several reasons why that number was not higher still. First, the cost was $200–$250 apiece, which was for many a prohibitively expensive price for an instrument which measured only one very specific physiological trait in one set of organisms. Second, without hands-on training, the dendrograph was difficult to assemble and use. Unaffiliated scientists had less success using the dendrograph than its inventor and his associates did.[25] And finally, MacDougal's possessive relationship to his instrument was problematic.

Scientists' interest in purchasing dendrographs was generated mainly through the circulation of Desert Lab scientific research papers in which it was featured. None of those who bought dendrographs ever published much dendrographic research of their own. When MacDougal retired, the instrument subsequently fell out of view. MacDougal's dendrograph was an early, successful foray into a technologically-mediated physiological ecology, showing the potential of technology to expand understanding of the interactions between organisms and their environment. However, the instrument itself was less significant than the conceptual shift it symbolized for ecologists.

The economic hardships of the 1930s caused the Carnegie Institution to curtail much of its funding for basic ecological research. By the late 1930s the Desert Laboratory's Tucson site was shuttered, and its buildings and grounds were transferred to the University of Arizona and the Forest Service in 1940. The Coastal Lab's equipment and personnel were moved up the coast to a facility on the campus of Stanford University. Sykes retired to a small home in the Arizona desert, continuing both his rambles and his construction projects well into his 80s. MacDougal officially retired as well, remaining in Carmel where he continued his dendrographic research, albeit at an increasingly leisurely pace. The methodological innovation of the Desert Lab held lasting impact. The unique instruments designed and used there showed other ecologists the potential of adapting laboratory instruments for the field. The dendrograph may not have become widely used itself, but it still served as an example of the potentials of precision measurement technology to expand the scope of ecology. Dendrograph research generated results that were not snapshots of plant communities, or even assemblages of series of snapshots, but instead, real-time visual charting of an organism's responses to environmental change.

The histories of the Desert Lab's instruments show the interconnectedness of research questions and instrument fabrication. Measurement capabilities boosted the standing of the emerging scientific approach of physiological ecology. The desert, with its sparsely-scattered plants striving against harsh environmental conditions, was the perfect place for studying plant physiology in the field. In the absence of dedicated instruments for that study, ecologists could not easily study the interactions between plants and their environments. Fabricating such instruments required collaboration and creative problem solving. Further, it required collaboration between instrument makers and scientists in an atmosphere of mutual purpose and singular focus. The Desert Lab provided both the collection of workers and the ecological setting that pushed the field in a new intellectual direction. In Tucson, the laboratory building sat in the middle of the scientists' field site of greatest interest, while the heat and aridity invited themselves inside the supposedly inviolable atmosphere of the lab. The physical barrier between the lab and field faded, and the intellectual barrier between fieldwork and labwork did as well. In a place encouraging collaboration across disciplinary boundaries, Godfrey Sykes and the Desert Lab scientists were capable of creating instruments of laboratory-style precision that were also adapted to the harshest field conditions.

Notes

1   My thanks to Alfred Nordmann, Astrid Schwarz, Seth Peabody, Nicole Seymour, and Heather McCrea for comments and suggestions, and to the Rachel Carson Center for Environment and Society, Munich, and the Max Planck Institute for the History of Science and the Max Planck Institute for the History of Science in Berlin for support.
2   "Historic American Buildings Survey: Desert Botanical Laboratory Complex, Photographs, Written Descriptive and Historical Data" HABS No. AZ-138 (Washington, D.C.: National Park Service, no date): 23.
3   Sykes was uncredited for some of his creations, like the long-period rain gauge, in research publications such as Livingston (1906). Sykes reclaimed credit for himself in his memoir, *A Westerly Trend*, and archival documents back up his claims.
4   See also Hornaday (1908); McGuire (2003); Nichols and Broyles (1997); Powell (1976); Broyles (2007).
5   MacDougal correspondence in Plant Biology 4/5/ "Desert Lab Administration 1905–1923 3 of 3," Carnegie Institution of Science archives (hereafter CIS).
6   Much written in the subdiscipline of laboratory studies proceeds from this assumption, as does Kohler (2002); see also Kohler (2012); Vetter (2011).
7   See Craig (2005); Good (1994); Cornell (2001); Kohler (1991); Ebert (1985); Clements (1960); Tobey (1981); see also *Centennial History of the Carnegie Institution of Washington,* Vols. 1–3, 5.
8   Note that this is less true of animal-centered studies than plant studies.
9   The term "physiological ecology" was not used by the Desert Lab staff, but serves us here to describe their methods, and to differentiate their work from forms of ecology. Frederic Clements referred to ecology as ideally becoming "field physiology," which is a useful and evocative term. See also the mid-20th century term "autecology" and the more contemporary "ecophysiology."
10   See also Cittadino (1990).
11   Original dendrographic records are found in folders 425–437, "Instrument Records and Summary," Box 30 and 31, D. T. MacDougal collection, Arizona Historical Society Archives, Tucson, AZ (hereafter AHS).

12 See folder 393, "Correspondence—Sykes, Godfrey and Glenton 1919" D. T. MacDougal collection, AHS.
13 MacDougal to Clements, n.d. ["Correspondence MacDougal, 1920," AZ 356, Box 8, University of Arizona Archives and Special Collections, Tucson, AZ].
14 U.S. Patent Office Patent Number 1385139, "Daniel Trembly MacDougal of Tucson, Arizona: Tree-Growth-Measuring Instrument," July 19, 1921; MacDougal to Merriam, March 28, 1924 [Plant Biology; folder "MacDougal, D. T. Dendrograph; 1924–1931 1 of 2," CIS].
15 Assignment of Patent [for dendrograph], Daniel Trembly MacDougal, April 23, 1924 [Plant Biology; folder "MacDougal, D. T. Dendrograph; 1924–1931 1 of 2," CIS].
16 MacDougal to Merriam, March 28, 1924; [unsigned] to MacDougal, April 5, 1924; Gilbert to Howe, April 8, 1924; Howe to Gilbert, April 9, 1924. All, [Plant Biology; folder "MacDougal, D. T. Dendrograph; 1924–1931 1 of 2," CIS].
17 MacDougal to Woodward, December 10, 1919, [file: "Plant Biology 4/5/ Desert Lab Administration 1905–1923 3 of 3," CIS].
18 MacDougal to W. M. Gilbert, March 6, 1925 [Plant Biology; folder "MacDougal, D. T. Dendrograph; 1924–1931 1 of 2," CIS].
19 ibid.
20 Correspondence [Plant Biology; folder "MacDougal, D. T. Dendrograph; 1924–1931 1 of 2" CIS].
21 MacDougal to Samuel T. Dana, February 14, 1927 [Plant Biology; folder "MacDougal, D. T. Dendrograph; 1924–1931 1 of 2," CIS].
22 MacDougal to to Sykes, June 8, 1926, numbers 1 and 2 [two letters] ["Sykes, Godfrey, 1926" AZ 356, Box 8, University of Arizona Archives and Special Collections].
23 Sykes to MacDougal, June 11, 1926 ["Sykes, Godfrey, 1926" AZ 356, Box 8, University of Arizona Archives and Special Collections].
24 The last dendrograph MacDougal made, which he created himself when he was retired, went to Zanzibar in 1947. It was a sad tale, as the instrument was either damaged in shipping or assembled incorrectly, and the buyer demanded a refund. The disappointment of the transactions was enough to make MacDougal decide never to make another dendrograph. [Plant Biology; folder "MacDougal, D. T. Dendrograph; 1931–1952 2 of 2," CIS] consists of nearly sixty pages of correspondence on this matter.
25 For examples of dendrographic incompetence see J. S. Karling to CIW, July 11, 1931 [Plant Biology; folder "MacDougal, D. T. Dendrograph; 1924–1931 1 of 2," CIS].

References

Bowers, J. E. 1992. *Plant World* and Its Metamorphosis from a Popular Journal into *Ecology*. *Bulletin of the Torrey Botanical Club* 119:*3*: 333–41.

Bowers, J. E. 1990. A Debt to the Future: Achievements of the Desert Laboratory, Tumamoc Hill, Tucson, Arizona. *Desert Plants* 10: 25–40.

Bowers, J. E. 1988. *A Sense of Place: The Life and Work of Forrest Shreve*. Tucson: University of Arizona Press.

Broyles, B. 2007. A Century of Science in the Pinacate. *Journal of the Southwest* 4:*2*: 129–34.

Cittadino, E. 1990. *Nature as the Laboratory: Darwinian Plant Ecology in the German Empire, 1880–1900*. New York: Cambridge University Press.

Clements, E. S. 1960. *Adventures in Ecology*. New York: Pageant Press.

Colville, F. V. and MacDougal, D. T. 1903. *Desert Botanical Laboratory of the Carnegie Institution*. Washington, D.C.: The Carnegie Institution.

Cornell, T. D. 2001. Carnegie Institution of Washington. In: *History of Science in the United States: An Encyclopedia*, edited by M. Rothenberg, 102–8, New York: Garland.

Craig, P. 2005. *Centennial History of the Carnegie Institution of Washington, vol. 4: The Department of Plant Biology*. New York: Cambridge University Press.

Ebert, J. D. 1985. Carnegie Institution of Washington and Marine Biology: Naples, Woods Hole, and Tortugas. *Biological Bulletin* 168: 172–82.

Ganong, W. F. 1904. The Cardinal Principles of Ecology. *Science* 19:482 (Mar. 25): 493–8.

Good, G. (ed.) 1994. *The Earth, the Heavens and the Carnegie Institution of Washington*. History of Geophysics, vol. 5. Washington, D.C.: American Geophysical Union.

Hornaday, W. T. 1908. *Campfires on Desert and Lava*. New York: Charles Scribner's Sons.

Kingsland, S. 2010. *The Evolution of American Ecology, 1890–2000*. Baltimore: Johns Hopkins University Press.

Kingsland, S. 1991. The Battling Botanist: Daniel Trembly MacDougal, Mutation Theory, and the Rise of Experimental Evolutionary Biology in America, 1900–1912. *Isis* 82:*3*: 479–509.

Kohler, R. E. 2012. Practice and Place in Twentieth-Century Field Biology: A Comment. *Journal of the History of Biology* 45:*4*: 579–86.

Kohler, R. E. 2002. *Landscapes and Labscapes: Exploring the Lab-Field Border in Ecology*. Chicago: University of Chicago Press.

Kohler, R. E. 1991. *Partners in Science: Foundations and Natural Scientists, 1900–1945*. Chicago: University of Chicago Press.

Livingston, B. E. 1906. *The Relation of Desert Plants to Soil Moisture and to Evaporation*. Washington, D.C.: Carnegie Institution of Washington.

Livingston, B. E. and Lawrence, D. B. 1948. Some Conversational Autobiographical Notes on Intellectual Experiences and Development: An Auto-Obituary. *Ecology* 29:*3*: 227–41.

MacIntosh, R. P. 1983. Pioneer Support for Ecology. *BioScience* 33:*2*: 107–12.

MacDougal, D. T. 1946. Streaming Movements of Solutions in Plants. *American Journal of Botany* 33:*5*: 318–28.

MacDougal, D. T. 1936. *Studies in Tree-Growth by the Dendrographic Method*. Washington D. C.: Carnegie Institution of Washington.

MacDougal, D. T. 1930. Lengthened Growth Periods and Continuous Growth. *Proceedings of the American Philosophical Society* 69:*1*: 329–45.

MacDougal, D. T. 1919. Growth in Organisms. *Science* (New Series) 49:*1278*: 599–605.

MacDougal, D. T. 1914. *The Salton Sea; a Study of the Geography, the Geology, the Floristics, and the Ecology of a Desert Basin*. Washington, D.C.: Carnegie Institution of Washington.

McGinnies, W. G. 1981. *Discovering the Desert*. Tucson: University of Arizona Press.

McGuire, T. R. 2003. The River, the Delta, and the Sea. *Journal of the Southwest* 45:*3*: 371–410.

Nichols, T. and Broyles, B. 1997. Afield with Desert Scientists. *Journal of the Southwest* 39: *3–4*: 353–70.

Nicolson, M. 1990. Henry Allan Gleason and the Individualistic Hypothesis: The Structure of a Botanist's Career. *Botanical Review* 56:*2*: 91–161.

Nicolson, M. and McIntosh, R. P. 2002. H. A. Gleason and the Individualistic Hypothesis Revisited. *Bulletin of the Ecological Society of America* 83:*2*: 133–42.

Powell, L. C. 1976. Hill of the Horned Toad. In: *From the Heartland: Profiles of People and Places of the Southwest and Beyond*, edited by L. C. Powell, 98–117, Flagstaff, AZ: Northland Press.

Shapin, S. 1989. The Invisible Technician. *American Scientist* 77: 554–63.

Sonnichsen, C. L. 1987. *Tucson: The Life and Times of an American City*. Norman: University of Oklahoma Press.

Sykes, G. 1945. *A Westerly Trend*. Tucson, AZ: Arizona Historical Society and The University of Arizona Press (1984).

Sykes, G. 1937. *The Colorado Delta*. Washington, D.C.: Carnegie Institution of Washington and American Geographical Society of New York.

Tobey, R. C. 1981. *Saving the Prairies: The Life Cycle of the Founding School of American Plant Ecology, 1895–1955*. Berkeley: University of California Press.

Vetter, J. 2011. Rocky Mountain High Science: Teaching, Research and Nature at Field Stations. In: *Knowing Global Environments: New Historical Perspectives on the Field Sciences*, edited by J. Vetter, 108–34, New Brunswick, NJ: Rutgers University Press.

Went, F. W. 1957. Fifty Years of Plant Physiology in the U.S.A. *American Journal of Botany* 44:*1*: 105–10.

Part IV
Calibrating Mind and World

14 Scientific Measurement as Cognitive Integration: The Role of Cognitive Integration in the Growth of Scientific Knowledge

Godfrey Guillaumin

Introduction

Among the various historical studies on the topic of "Scientific Revolution," there is not sufficient consideration of the deep cognitive changes generated by scientific measurements, although there is usually an emphasis on the theory change.[1] A well-known case of theory change is Nicholas Copernicus's (1473–1543) planetary theory, which was stated as an alternative explanation to Claudius Ptolemy's (*ca.* AD 100–*ca.* AD 170). Developing a planetary theory supposing that the earth rotates around a motionless sun was not only a simpler theory but also a stronger explanatory theory. However, an interesting methodological and cognitive issue was that Copernicus's astronomical measurements and data were not notably different from Ptolemy's. In fact, in many cases, Copernican data concerning key astronomical parameters was the same as in Ptolemy's theory. Therefore, in this instance, we have two inconsistent theories concerning planetary motion derived from nearly the same measurements.

However, the methodological and cognitive relationship between Copernicus's planetary theory and Johannes Kepler's (1571–1630) was highly different. Kepler was a Copernican astronomer because he maintained the theory about a stationary sun, but he was not Copernican when conceiving, developing, and performing astronomical measurements. Consequently, we could say that between Copernicus and Kepler, there was not a theory change but a measurement (methodological and cognitive) change, i.e., there was a cognitive change without a theory change. Elsewhere, I have analyzed this situation and called Kepler's measurement achievements a "cognitive integration" (see Guillaumin 2012). The aim of this contribution is to show that "cognitive integration" is a type of change that is more fundamental and deeper than theory change.

Copernicus's Astronomy with Ptolemaic Measurements

Ptolemy's astronomy and Copernicus's were not only two different theories for explaining planetary motion but also generated scientific revolutions:

"The Copernican Revolution broadly understood has all of these meanings, astronomical, scientific, and philosophical, and must therefore be studied as both scientific and intellectual history" (Swerdlow 2004: 76). Nevertheless, the change from the Ptolemaic to the Copernican planetary theory was a central part of this revolution, in which Copernicus's planetary theory achieved a more powerful and simpler explanation than Ptolemy's. According to Swerdlow, there were 27 different planetary issues that Copernicus explained without using arbitrary assumptions, as Ptolemy's theory did. A number of these explanations were the following: (1) planetary retrogradation; (2) the correspondence of the heliocentric periods of the earth and the planets with their order from the central sun; (3) a definition of the planets' synodic periods—the periods between successive conjunctions with the sun; (4) why the radii of the epicycles of superior planets are always parallel to the direction from the earth to the mean sun; (5) why the superior planets all move in the same direction on their epicycle; (6) why the centers of the epicycles of the inferior planets always lie in the direction of the mean sun; (7) why superior planets reach opposition to the sun; (8) why inferior planets reach only a limited elongation; (9) why superior planets are retrograde on either side of opposition; (10) why inferior planets are retrograde on either side of inferior conjunction; (11) why superior planets are at the perigee of their epicycle when at opposition; (12) why inferior planets are at perigee when at inferior conjunction; (13) why superior planets are brightest at opposition; (14) why inferior planets are brightest near inferior conjunction; (15) why planets more distant from the earth have smaller epicycles than planets closer to the earth; and so on. Copernicus's planetary theory revealed a much more explanatory capacity than Ptolemy's, mainly because a *heliocentric theory explains all the characteristics of the geocentric theory, but not the other way around* (Swerdlow 2004: 89, emphasis in original).

However, what convinced Copernicus about his planetary theory was not accuracy, for he knew that the same parameters would produce the same apparent motions in either system. To a few astronomers of that epoch, "the initial choice between Copernicus's system and Ptolemy's could only be a matter of taste, and matters of taste are the most difficult of all to define or debate" (Kuhn 1957: 172). Ptolemy's astronomy and Copernicus's were relatively the same in (unsophisticated) precision, the mathematics used in their models, and the instruments for observation. It can be stated that: "Copernican astronomy was neither simpler nor more accurate than Ptolemaic astronomy" (Barker and Bernard 1988: 302). The revolutionary element of Copernicus was to place the earth moving and the sun still, not in the center, as suggested by Aristarchus of Samos (*ca.* 310 BC–*ca.* 230 BC), but "in most respects his *De revolutionibus* (1543) follows Ptolemy's *Almagest* so closely that he can equally well be regarded as the last great practitioner of ancient astronomy" (Thoren 2003: 3). Copernicus inherited the astronomical data and observations made using the standards and instruments of his time.

He did not develop new measurement tools, new mathematics to analyze data or new measurement techniques.

An example can illuminate this point further. Ptolemy did not determine the true distances of the planets in *Almagest*, but in *Planetary Hypotheses* he accounted for them by the theory of nested spheres.[2] Copernicus's planetary theory established for the first time the true distances of the planets in terms of the geocentric distance of the sun, but surprisingly, he did not utilize the solar distance. To establish true planetary distances, it was necessary for Copernicus to use a geometric method developed and used by Hipparchus of Nicaea (*ca*. 190 BC–*ca*.120 BC) and Ptolemy, called an "eclipse diagram." This method used, among other parameters, the lunar and solar parallaxes. In Ptolemy's version of this method, the lunar distance and the lunar model were essential features of the determination of solar distance. Copernicus used Ptolemy's method, but his lunar model was specifically designed to correct the variation in the distances and the parallax generated by Ptolemy, successfully preserving the quality of the Ptolemaic predictions. Nevertheless, despite the fact that it appears that Copernicus obtained original values for the apparent solar and lunar diameters and for the ratio of the lunar and shadow radii (which were fundamental parameters in the "eclipse diagram"), there are good reasons to think that he merely adjusted Ptolemy's parameters.

After a meticulous study of Copernicus's planetary calculations, Janice Henderson has concluded that Copernicus "has merely adjusted the values of Ptolemy and al-Battani to fit the lunar distances which arise from his new lunar model. ... Both al-Battani and Copernicus relied heavily on Ptolemy, and I would venture to say that they wished to obtain a solar distance which differed but little from that of Ptolemy" (Henderson 1991: 8). Furthermore, Copernicus not only adhered to the subject matter of the *Almagest*, presented in roughly the same order, but his mathematical methods, used in the derivation of parameters and the construction of astronomical tables, are virtually identical to Ptolemy's: "The determination of the solar distance is but one example of Copernicus's use of the actual numerical values in the *Almagest*. The underlying concern throughout *De Revolutionibus* is to show that the Copernican models produce results which agree with the Ptolemaic models" (ibid.: 9–10). So, here we have two inconsistent planetary theories, one of them with a great explanatory capacity, but constructed with the same measurement methods and data as the other.[3]

For the accuracy of astronomical measurements, from ancient times until Copernicus, van Helden argues that we must draw a distinction between *accuracy* and *convenience* (see van Helden 1983: 54). According to John North, the astrolabe was undoubtedly useful in practice to know the time, despite its imprecision (see North 1976). It can be said that including ancient astronomy until Copernicus, data were more convenient than accurate. For example, the accuracy of angular distances in Copernicus's measurements has been estimated between 1/8° and 1/10° (6'), which was almost 10 times better than the accuracy of medieval astrolabes but still not sufficient to detect parallax. Without the development of new methods and measurement tools, Copernican theory would have remained a mere mathematical hypothesis.

Kepler's Copernican Astronomy without Copernican Measurements

Scientific measurement, not as a single act but as an historical development, is a very complex dynamical process. It requires a *dynamic integration* for its development (in terms of coupling and self-adjustment) of three different types of cognitive resources that reciprocally evolve through history: namely, conceptual, mathematical, and instrumental resources. A few years after *De Revolutionibus* was published, astronomy deeply transformed and integrated these three measurement resources through a *dynamic integration*, generating unprecedented cognitive measurement tools.[4] Instrumental transformations could generate conceptual and/or mathematical transformations, and vice versa, and one of the main reasons is the following. Measuring instruments work as data providers that are usually beyond our natural system of perception. To be *reliable*, suppliers are required to transform and refine them, or otherwise, the data could have significant deviations when making calculations. Second, we cognitively *think based on instruments*, which means that certain ideas were conceived through the instruments we use, and we could hardly conceive that part of the world without them. In that sense, instruments in general, but measuring instruments in particular, are an extension of our cognition and minds.[5]

Based on these considerations, it can be argued that performing measurements to investigate the empirical world requires the mathematical integration and instrumental modification of specific conceptions about the world. Such integration can take decades and necessarily requires the transformation, creation or refinement of empirical concepts, mathematical concepts (or systems) and instruments (or experiments). It is mandatory to achieve the integration required to fit these three elements mutually. Because the idea of *integration* is at the core of what is required to make appropriate, relevant and correct measurements, I consider it appropriate to call this theory of making measurements the *integrative theory of scientific measurement*.

During the late 16th century, an integrative measurement process arose in astronomy. Astronomy has always (and in virtually all cultures around the world) used instruments to establish certain astronomical parameters, so Hipparchus, Ptolemy, and Copernicus, just to mention a few astronomers, used and developed measurement instruments. The technological novelty during the second half of the 16th century exhibited two aspects: the first was the consistently increased precision of existing astronomical instruments to a level never observed before, and consequently, new instruments were invented. The second was based on integrating these previously unknown tools with brand new concepts and mathematics. Let us examine this briefly.

In the early 17th century, the terms "instruments" and "experiments" did not mean what they do now. There is one important difference that has disappeared over time but is crucial for historical understanding, as it depends largely on how instruments have transformed measurement practices and empirical knowledge. Jim Bennett states that: "from the seventeenth century instruments were conventionally classified into mathematical, optical and natural philosophical types" (Bennett 1983: 105). The mathematical instruments "such as the dioptra and the astrolabe—and yes, even Galileo's proportional compass—were part of a long tradition, certified by usage and custom in the restricted realm of the mathematical subjects.[6] The *dioptra* was based on the well-known principles of Euclidean and Ptolemaic optics; the astrolabe embodied the technique of mathematical projection of a sphere onto a plane; and the proportional compass was based on the theory of proportions going back to Eudoxus" (van Helden 1994: 9). Hipparchus improved the *dioptra* and created the astrolabe, and he attributed to Eratosthenes of Cyrene (*ca.* 276 BC–*ca.* 194 BC) the creation of the armillary sphere (see Evans 1998: 75). Greek positional astronomy used some "mathematical instruments" and incorporated geometry and trigonometry into their models of the cosmos, as well as records of celestial observations collected for more than five centuries by the Babylonians and the Egyptians. However:

> from the sources we now have in translation, it appears that the very accurate parameters of the motions of the planets, the sun, and the moon current the Mesopotamians after approximately 700 B.C. were the result not of accurate measurements but rather of very mediocre measurements continued over a long period of time. We must go to late antiquity, the period from Hipparchus to Ptolemy, to find the first attempts at accurate and convenient measurement to solve specific astronomical problems. Even in the case of Ptolemy, however, it is obvious that he relied heavily on Mesopotamian sources.
>
> (van Helden 1983: 53)

Although the *Almagest* was considered to be the most complete astronomical model for over 1,500 years, as well as a useful and true synthesis of contemporary astronomical knowledge, we must remember that: "although Ptolemy's stated observations are not particularly good and in one case err by more than a degree, they do match his tables in an uncanny way. In fact, this situation prevails throughout the entire *Almagest*, which leads to the suspicion that something is going on that does not meet the eye ..." (Gingerich 1997: 17). The Almagest was not conceived as a research book in our sense, but more as a textbook that compiled the astronomical knowledge of the period influenced by Euclidean geometry.

Owen Gingerich explains the proper correspondence, when it exists, between Ptolemy's models with the data used in the following way: "I am inclined to think that ... he may no longer have felt in necessary to find perfect observations, for he could simply have used his theory to judge how much observational error an observation contained, and he could correct it accordingly" (ibid.). Much has been written about whether Ptolemy falsified data, but it is only necessary to emphasize that such falsification would have been committed (unnecessarily) to develop a systematic program of detailed and precise measurements of celestial parameters. In brief, there were measurements in ancient astronomy integrating geometry, instruments and concepts, but qualitatively, these resources had reached their cognitive limit, as they had reached their maximum development. One of these three had to reach another level of development to perform other types of measurements: in the case of astronomy, it was the instruments.

As we have observed, Copernicus made far-reaching conceptual and explanatory changes in astronomy with his planetary theory. What is usually forgotten in the history of astronomy is that without the transformation of the *instruments* and *mathematics*, his model would not have become widely accepted. Tycho Brahe (1546–1601) systematically developed a program, unprecedented in the history of astronomy, for more accurate astronomical measurements, and the mathematical mind of Johannes Kepler finally *integrated* the new measures on the cosmos using mathematical transformations.

In early 1570, Tycho recognized the need to build "a new astronomy based entirely on logic and mathematics, *without recourse to any hypothesis.* [He] agreed on the need for new and accurate observations *before attempting to explain* the celestial motions, and it is obvious that Tycho was aware of the need for good instruments to obtain those observations" (Hellman 1974: 402, emphasis added). Between 1576 and 1596, he had collected a large amount of data, obtained with instruments that he had ordered,[7] so precise that they made all astronomical observations from before the 16th century obsolete, including those used by Copernicus (see Blair 1990). Estimations have shown that the range of accuracy of Tycho's instruments was between 30" to 50," or approximately 10 to 20 times more than those used by Copernicus (see Chapman 1983: 135). Accurate data depended not only on the instrument:

> The accuracy of the observations depended on the instruments and the care with which they were used. Although Tycho's were without magnification, error was minimized by their huge size and by the graduations carefully marked on them to facilitate angular measurements on the celestial sphere, altitudes, and azimuths. *Tycho checked instruments against each other and corrected for instrumental errors* ... He observed regularly and achieved an accuracy within a fraction of a minute of arc, an accuracy unsurpassed from the time of Hipparchus to the invention of the telescope.
>
> (Hellman 1974: 405, emphasis added)

To make precise measurements, some methodological considerations must be addressed. In his instruments, Tycho made two major technical advances: diagonal scales used in the arms of reading instruments, which allowed him to measure fractions of a degree without increasing the size of the observation instruments, and improving the instruments through cracks in the alidade in sextants or quadrants, which decreased alignment errors. However, the major methodological development was an improved data collection method, in which he repeated observations to obtain more data for each element observed.[8] "Tycho had an unprecedented concern for accuracy and a unique ability to detect and eliminate flaws in either observational procedure or mechanical design" (Thoren 2003: 12), and he had transformed the art of making good astronomical observations, which was the dominant mode of astronomical practices from the ancient astronomers to Copernicus, into a high-precision astronomical *science*. After Tycho, astronomical observations acquired new precision and methodological and cognitive standards, indicating that he crossed a "cognitive threshold" in observational astronomy that had never been so high in the history of astronomy.[9]

Although Tycho's observations achieved accuracy, he lacked the proper understanding about the optical elements of observations, including the causes and effects of refraction. Tycho was seriously considering refraction, and he even argued that Copernicus had his own latitude incorrect by over two minutes of arc due to ignoring the effects of refraction.[10] However, Tycho himself did not properly understand the theory of refraction: "it is needless to say that the accuracy cannot be so great as Tycho fondly hoped, as the errors of observation would be increased by neglect of refraction and by his ignorance of the existence of aberration and nutation" (Dreyer 1890: 351).

Tycho did not theoretically understand the effects caused by refraction in astronomical measurements, but Kepler did. Moreover, although the accumulation of data with unprecedented accuracy was a remarkable achievement in itself, it did not estimate the cognitive, methodological and epistemic consequences required. Specifically, it lacked the correct ratios and proportions "hidden" among such data in the framework of a *theory of the heliostatic cosmos*, for which it was insufficient to continue adding data. It required a conceptual and mathematical analysis of these relations and proportions. Kepler made that analysis about different physical and astronomical phenomena: one of them is refraction and the other is the motion of Mars.

Kepler met Tycho in early 1600 and slowly began to gather the data that Tycho had accumulated. Kepler made an early conceptual development on refraction from Tycho's data. In his measurements, Tycho recognized that the sun apparently moved upward as it approached the horizon. He thought that this phenomenon was due to the refraction of light near the horizon and devoted a series of systematic measurements of the positions of the sun, moon and fixed stars. Comparing these measurements with predictions, he concluded that refraction was not detected above 45 degrees in the cases of the sun and the moon, or 20 degrees for fixed stars and planets. Tycho's measurement data were incorporated into three different refraction tables, one for the sun, another for the moon and the third for the fixed stars and planets (see van Helden 1983: 57–8).

Data from the three tables were accurate but not joined conceptually or mathematically by Tycho. Kepler tried to remedy the situation of having three different refraction tables, and he attempted to integrate these data in *Paralipomena ad Vitellionem. Astronomia pars optica* (1604). He considered certain optical principles and argued that refraction is due to differences in the densities of the ether and the atmosphere. He sought a mathematical relationship based on these principles that agreed with all measurements made from Tycho's three tables. Kepler was able to find a formula or law of refraction, which gave good results for calculating latitudes higher than ten degrees, although it had some errors from our perspective. With that mathematical relationship, Kepler built a single refraction table for all celestial objects, even more accurate than Tycho's. Kepler's analysis of Tycho's data joined measurements with mathematical and theoretical considerations to develop a law of refraction.[11] Tycho had accurate data, but no mathematical or conceptual considerations for finding a mathematical proportion to the refraction.

Another cognitive problem existed with refraction and, in general, with the astronomical observations, which Kepler faced and solved. Astronomers measured the apparent distances between the fixed stars, planets and the edges of the sun and the moon with instruments and expressed those measurements in arcs of visual angles (*anguli visorii*). Human vision is not only involved in the measurements but is also a component of them, so "this whole enterprise in astronomy rests upon optical reasons" (Kepler 2000 [1604]: 321). This describes Kepler's consideration of the importance of the study of optics in two ways: a study of the nature of light as well as of the functioning of human vision. Despite its importance for generating reliable, accurate knowledge, this last topic has not been adequately evaluated in the development of astronomy.[12]

Before any consideration, we must understand how Kepler conceived the astronomical practice. In the preface to his *Paralipomena ad Vitellionem. Astronomia pars optica (Additional Aspects of Witelo. The optical part of astronomy)*,[13] usually known simply as *Optics*, Kepler explains that for him, astronomy aims to study the *motion* of celestial bodies. This study has two parts: the first is the investigation of the *shapes* of the movements of the celestial bodies, and the second, derived from the first, is to investigate the *positions* of celestial bodies at any given time, for the practical purpose of prediction.

The first part is formed by geometry; the second by arithmetic. Three components comprise the first geometric part: the first one was called *mechanic* and is related to using instruments for astronomical observations; the second was called *historical* and contains voluminous records to understand these observations; the third was called the *optical* component, and it involved the optical principles that include both the nature of light and the functioning of human vision. Kepler recognized that Tycho had greatly advanced the *mechanical* and *historical* aspects, both for the construction of instruments and the

"24 books of the most meticulous observations of this sort, embracing the past 40 years or so" (Kepler 2000 [1604]: 13). Nevertheless, the relationship of *optics* and astronomy was not only neglected but also completely misunderstood. Kepler's *Optics* can be considered a theoretical reform of optical knowledge and the complement to astronomy required to achieve greater reliability in the data generated by measurements.

The main motivation for Kepler to develop a detailed analysis of optics was to correct the optical knowledge that he inherited and was linked to measuring fundamental astronomical parameters. Astronomical practices depended entirely on optical reasoning, and the only way to guarantee a reliable link between the mental construct and the physical reality of the heavenly bodies was to understand theoretically both the nature of light and the functioning of the human eye. Kepler thus argued that "supposing the place of the celestial body to be known with complete precision, throws the demonstration into difficulty: the nature of light, beset by the inconstancy of optical causes, does not always allow such precision of instruments" (ibid.: 6).

One of the most common astronomical measurement methods since ancient times was what Kepler called the "theorem." This was "to use a compass to measure the magnitudes of solar eclipses, the ratios of the diameters of the sun and moon, and the inclinations to the vertical of the circle drawn through the centers of the luminaries" (ibid.: 57). However, Kepler said that astronomers who used that method previously thought (incorrectly) that they were "avoiding the inadequacy of the eyes, and avoiding the error which generally occurs in a bare estimation" (ibid.). This assumption was one of the biggest mistakes made by astronomers and had contaminated many astronomical records. This unfortunate methodological situation prevented the realisation of precise astronomical measurements, so Kepler states the following:

> It is indeed well worthwhile here to see how much detriment would result from the ignorance of the proof of this theorem [method]. For since it escaped a number of authors, the result was that in believing in the theorem without restrictions they fell into a large error. For however many eclipses were observed in this way, they all had come out much greater in the sky than it appeared in the ray: all showed a much greater lunar diameter in the sky than in the ray ... *It is my hope in these pages to remove these considerable difficulties* ...
>
> (ibid.: emphasis added)

Kepler's reform in optics and astronomy, which now are considered as two different areas, has one main pillar: his concern for the theoretical and methodological aspects of measuring high-precision astronomical data. From Kepler's perspective, a reform was needed in both astronomy and optics.[14]

Such a reform should be performed from the core. One of the most routine, everyday and "simple" astronomy activities was to identify the position of the planets, and yet it presented significant difficulties, in that: "because of the eccentric, the planets appear either slow or fast. The cause is partly physical, partly optical. The physical part of the cause does not give the sense of vision a reason for error but also represents to the vision that which in fact occurs, [an account] of which is in the *Commentaries on the motions of Mars* [this book was later known as *Astronomia Nova*]" (ibid.: 321).

Conversely, Kepler warned astronomers that they should not rely on their sense of sight while making measurements of the apparent diameter of the full moon and sun (see ibid.: 298). Unlike Tycho Brahe, Kepler was convinced that because of large variations in the visual capacity of each individual observer, it was impossible to establish accurate astronomical tables. Highly developed and accurate observation instruments were required, but light behaviour must be first understood. Kepler said that "it cannot ... be argued from this accident of the sense of sight to what happens outside of consideration of the sense of sight, nor can tables be established for the sake of the sense of sight, which represent neither the object itself nor the defects of all senses of sight. For the astronomer should not present anything other than those things that in actual fact occur. The sense of vision, however, we leave to the physicians to remedy" (ibid.).

Even the traditional instruments used for these observations were insufficient. He realized that, using a pinhole camera as an instrument to observe and make measurements of the sun's apparent diameter, the image simply did not show the true diameter of the sun or the moon, and the astronomer was easily fooled. Before using or redesigning the pinhole camera (as Kepler eventually did), a better understanding of the causes of these optical projections was required "to teach how to enter into a most certain procedure for measuring the quantities of eclipses" (ibid.). One method used by Kepler to investigate the varying distance between the earth and the sun was to measure the apparent diameter of the sun over a year; although this method yielded inconclusive results, it provided the opportunity to complete astronomical measurements with optical theory.[15]

Conclusion. "Cognitive Integration" as a Deeper Change than a Theory Change

Copernicus elaborated a new and more explanatory theory of planetary motion than Ptolemy's with almost the same astronomical data and measurements as those used by Ptolemy. This great theory change was at the explanatory level but not at the cognitive level, mainly because there were not changes at the measurement level. Scientific measurement, considered in its historical development, is a complex and dynamical process that integrates three different aspects: mathematics, concepts and instruments.

As we have observed, Kepler's planetary measurements were a deep cognitive change, compared with Copernicus's, that involve more accurate data, physiological considerations related to optical instruments, integration of the physical part of astronomy with the mathematical aspects, cross-checking of observational errors and convergence of data with a very high instrumental precision and a very high mathematical accuracy. Kepler integrated all of these dissimilar and diverse aspects into specific astronomical measurements, some of which are known today as "Kepler's planetary laws." This specific historical case clearly demonstrates that an only explanatory advance is insufficient to the growth of scientific knowledge; it also requires a cognitive change, which I have called "cognitive integration," conducted by the specific process of scientific measurement.

Notes

1 An exception to this statement is Kuhn (1961: 31–63).
2 This theory assumes crystalline and nested spheres bear planets and move them around the earth. The least distance of a planet is equal to the greatest distance of the planet just below it, and that the ratio of the relative distances of a planet is equal to the ratio to the ratio of the true distances.
3 To see in detail in which sense the data of Ptolemy and Copernicus are convenient rather than accurate, see Newton (1977), Evans (1998), and Gingerich (1997).
4 It is interesting that David Sherry (2011) detected the same phenomena that I am naming "dynamic integration" in the case of measurement of heat: "Unlike his predecessors, [Joseph] Black used reading from his device to achieve a theoretical understanding of thermal phenomena by applying mathematics to the readings. Surprisingly, Black's mathematical techniques emerged long before the thermoscope. He *melded* this older, conceptual tradition with the experimental tradition that immediately preceded his discoveries" (ibid.: 512, emphasis added).
5 Bertoloni (2006) shows how cognitively speaking, some objects, instruments or experiments used in specific ways functioned as extensions of the human mind to *conceive* new ideas and sometimes to "modify experimentally" speculation.
6 The main tools developed and / or used during the 17th century natural philosophy and astronomy were the vacuum pump, the astrolabe, the barometer, calculating machines, the compass, the cross geometry, the microscope, the compass of proportion quadrant, the slide rule, the telescope, the thermoscope and pendulum clock. For development of mathematical instruments during this epoch see especially Bennett (2011: 697–705).
7 The instruments that Tycho ordered to be constructed and he used in the observatory on the island of Hven were the following: a metal azimuthal quadrant (1576) was one of the first instruments built in Hven. Tycho used it to observe the comet in 1577 and it has an estimated accuracy of 48.8 arc sec. A Calculating Globe (1580), which Tycho used to record the positions of stars that in 1595 summed up to a thousand and which also served to locate the azimuth coordinates and turn them into conventional celestial coordinates. An Armillary Sphere of 1.6 meters of diameter (1581). A triangular Sextant (1582) of 1.6 meters of radio. Armillary Great Circle (1585) three feet in diameter and whose accuracy was estimated at 38.6 arc seconds. A wooden mobile Quadrant (1586) of 1.6 meters radio and whose accuracy was estimated at 32.3 seconds of arc. A mobile metal Quadrant (1588) whose accuracy was estimated at 36.3 seconds of arc.
8 For details see Thoren (1973: 25–45) and Wesley (1978: 42–53).

9 To get an idea of the cognitive threshold, it will be useful to consider the following: The angular diameter of the moon on average is about 30 minutes of arc (half a degree). The average estimated accuracy of the instruments of Tycho is about 30 seconds of arc, while in Copernicus's measurements the accuracy was 6 minutes. The precision reached by Tycho is out of range of that achieved not only at first glance but also of the astronomical instruments used before him apart from using methodological considerations to eliminate observation errors.

10 Tycho recognized the serious effects of refraction because they interfere with three very important astronomical parameters: (1) they affected the correct setting of the latitude of his observatory, (2) they introduced a radical change in the value that Tycho and everyone before him had been using to determine the obliquity of the ecliptic, and (3) they affected the establishment of the solar parallax (see Thoren 2007: 226).

11 A detailed study of the methodological considerations that Kepler used in the construction of his law of refraction is in Buchdahl (1972).

12 As it has been shown by Hon and Zik (2009).

13 Donahue (1994) and Lindberg (1996) maintain that Kepler reached, in that book, the highest theoretical peak of perspective tradition in optics, which begins with the Greeks and passed by the Arabs as Alhazen (965–1040) and medieval authors as Witelo (1230–ca. 1300) of the 13th and 14th centuries. At the same time, Kepler opens the modern approach to optics, which topics are to understand three phenomena: how the eyes work, how refraction works and a new level of understanding of how lenses form images.

14 In the context of the history of astronomical measurements, from ancient times to Kepler's, it is important to trace a distinction between "precision" and "accuracy." *Precision* refers to the degree of refinement with which an operation with *astronomical instruments* is performed or a measurement stated, and it require very different aspects from manual skills to optical considerations, from practical knowledge to understanding mechanical devices, etc. *Accuracy*, on the other hand, is the degree of conformity to some recognized standard (typically mathematical) value or to mathematically arrive to a true value of an astronomical parameter through a mathematical (geometric) procedure (see Lloyd 1991: 304). Mathematical procedures of measurement in astronomy usually obtained accuracy for different astronomical parameters, but astronomical instruments were usually underdeveloped for measuring some of these parameters. Kepler's optical and technical considerations about astronomical instruments managed to bring *accuracy* with *precision*. This is one of the key traits of *cognitive integration* of the scientific measurement.

15 In the last chapter of his *Optics*, chapter 11, Kepler offers the construction details of an observation instrument built by him, and he incorporated the theoretical reforms on optics he had developed in the previous chapters. Such an instrument was called an "instrument of ecliptic" and consisted of the combination of a pinhole camera (which very precisely measured the apparent diameter of the sun and the moon) with a compass, traditionally used to establish the positions of astronomical bodies. Among other benefits, such an instrument was able to measure the image of the diameter of the sun projected on a screen with the appropriate theoretical adjustments derived from his principles of optics, and it avoided the eye's natural distortions that are different in each observer (see Kepler 2000 [1604]: 297). A detailed study of the problems, both theoretical and material, that Kepler had to overcome in building the "instrument of ecliptic" is that of G. Hon and Y. Zik, *Kepler's Optical Part of Astronomy (1604): Introducing the Ecliptic Instrument*.

References

Barker, P. and Bernard, R. G. 1988. The Role of Comets in the Copernican Revolution. *Studies in History and Philosophy of Science Part A* 19:*3*: 299–319.

Bennett, J. A. 2011. Early Modern Mathematical Instruments. *Isis 102*: 697–705.

Bennett, J. A. 1983. A Viol of Water or a Wedge of Glass. In: *The Uses of Experiment*, edited by D. Gooding, T. Pinch, and S. Schaffer, 105–14, Cambridge: Cambridge University Press.

Bertoloni, D. 2006. *Thinking with Objects. The Transformation of Mechanics in the Seventeenth Century*. Baltimore and London: The Johns Hopkins University Press.

Blair, A. 1990. Tycho Brahe's Critique of Copernicus and the Copernican System. *Journal of the History of Ideas* 51: 355–77.

Buchdahl, G. 1972. Methodological Aspects of Kepler's Theory of Refraction. *Studies in History and Philosophy of Science Part A* 3:*3*: 265–98.

Chapman, A. 1983. The Accuracy of Angular Measuring Instruments Used in Astronomy Between 1500 and 1850. *Journal for the History of Astronomy* 14: 133–7.

Donahue, W. H. 1996. Kepler's Invention of the Second Planetary Law. *British Journal for the History of Science* 27:*1*: 89–102.

Dreyer, J. L. E. 1890. *Tycho Brahe: A Picture of Scientific Life and Work in the Sixteenth Century*. Edinburgh: Adam & Charles Black.

Evans, J. 1998. *The History and Practice of Ancient Astronomy*. Oxford: Oxford University Press.

Gingerich, O. 1997. *The Eye of Heaven: Ptolemy, Copernicus, Kepler*. Netherlands: Springer.

Guillaumin, G. 2012. De las cualidades a las magnitudes: la medición científica como integración cognitiva en el surgimiento de la astronomía moderna. *Signos Filosóficos* 14:*28* 57–89.

Hellman, D. C. 1974. Brahe, Tycho. In: *Dictionary of Scientific Biography*, vol. 2, edited by C. Gillispie. Detroit: Charles Scribner's Sons.

Henderson, J. 1991. *On the Distances between Sun, Moon, and Earth According to Ptolemy, Copernicus, and Reinhold*. Leiden: Brill Academic Publishing.

Hon, G. and Zik, Y. 2009. Kepler's Optical Part of Astronomy (1604): Introducing the Ecliptic Instrument. *Perspectives on Science* 17:*3*: 307–45.

Kepler, J. 2000 [1604]. *Optics*. New Mexico: Green Lion Press.

Kuhn, T. S. 1961. The Function of Measurement in Modern Physical Science. In: *Quantification. A History of the Meaning of Measurement in the Natural and Social Sciences*, edited by H. Woolf, 31–63, Indianapolis: Bobbs-Merrill Co.

Kuhn, T. S. 1957. *The Copernican Revolution. Planetary Astronomy in the Development of Western Thought*. Cambridge, Mass.: Harvard University Press.

Lindberg, D. C. 1996. *Theories of Vision from Al-Kindi to Kepler*. Chicago: University of Chicago Press.

Lloyd, G. E. R. 1991. *Methods and Problems in Greek Science*. Cambridge: Cambridge University Press.

Newton, R. 1977. *The Crime of Claudius Ptolemy*. Baltimore and London: The Johns Hopkins University Press.

North, J. D. 1974. The Astrolabe. *Scientific American* 230:*1*: 96–106.

Sherry, D. 2011. Thermoscopes, Thermometers, and the Foundations of Measurement. *Studies in History and Philosophy of Science* 42: 509–24.

202 *Reasoning in Measurement*

Swerdlow, N. M. 2004. An Essay on Thomas Kuhn's First Scientific Revolution, The Copernican Revolution. *Proceedings of the American Philosophical Society* 148:*1*: 64–120.

Thoren, V. 2007. *The Lord of Uraniborg: A Biography of Tycho Brahe.* Cambridge: Cambridge University Press.

Thoren, V. 2003. Tycho Brahe. In: *Planetary Astronomy from Renaissance to the Rise of Astrophysics, Part A, Tycho Brahe to Newton*, edited by R. Taton and C. Wilson, 3–21, Cambridge: Cambridge University Press.

Thoren, V. 1973. New Light on Tycho's Instruments. *Journal for the History of Astronomy* 4: 25–45.

van Helden, A. 1994. Telescopes and Authority from Galileo to Cassini. *Osiris* 9: 9–29.

van Helden, A. 1983. The Birth of the Modern Scientific Instrument, 1550–1700. In: *The Uses of Science in the Age of Newton*, edited by J. Burke, 49–84, California: University of California Press.

Wesley, W. 1978. The Accuracy of Tycho Brahe's Instruments. *Journal for the History of Astronomy* 9: 42–53.

15 Measurements in the Engineering Sciences: An Epistemology of Producing Knowledge of Physical Phenomena

Mieke Boon

Introduction

One of the earliest uses of measurements was their well-known role in trade, where the ability to use rudimentary measures of weight not only made it possible to barter with food and raw materials but also enabled things to be built and manufactured. The simple ability to measure the length of things by means of specific units, combined with some elementary arithmetic and geometry, enabled craftsmen to design and construct things such as cathedrals, castles, bridges, houses, musical instruments, furniture, tools, and clothes. Reflecting further upon this observation, we come to realize that it is the ability of humans to measure and apply basic mathematics that *makes it possible to design things at all*. Designers can work out on paper or by means of computer simulations how to build something that does not yet exist—how to construct, say, a building or a ship whose size matches our needs and is stable and strong enough, while also satisfying our aesthetic ideals. More than this, though, our ability to measure and calculate makes subsequent *epistemic uses* of the design possible. In the actual process of construction, the epistemic uses of a design include, for instance, calculating the quantity of materials to be used and the dimensions of the component parts.

This perspective on the role of measurements and mathematics in the design of artefacts can be extended to the design of more advanced technologies, such as those found in chemical engineering, biomedical engineering, and nanotechnology. These differ from the examples of technological artefacts just mentioned in that the latter are considered primarily in terms of having a function suitable for certain uses by humans (see also Houkes and Vermaas 2010), whereas more advanced technologies usually "do something" themselves: they produce something, they generate changes and transformations and they perform technological activities (or have the capacity to do so). This kind of technological functioning is often described in terms of physical-technological processes (e.g. the conversion of chemical compounds or the conversion of light into an electric current) and capacities (e.g. the capacity of a material to resist an electric current or the capacity of a chemical catalyst to accelerate a chemical reaction).

As a consequence of this difference, the "naïve" picture sketched above, in which the ability to measure properties of objects such as size, shape, and weight enables the design of, say, a house, is insufficient for understanding how measurements enable the design of advanced technologies.[1] This is because such design additionally involves the measurement of physical properties that manifest only in specific physical-technological circumstances. Crucial to the argument developed in this contribution is the fact that these kinds of properties can be measured only when they manifest, i.e. when they become apparent as a result of specific physical-technological circumstances; these properties are "capacities," so to speak. Strictly speaking, then, physical properties are actually measured by means of the measurement of *physical phenomena*.[2, 3]

The key idea being proposed here is that the ability to measure the physical properties of materials and technological devices is the very thing that makes the design of a technology possible in the first place and enables the epistemic uses of a design in the actual manufacture of a technology.[4] The reasoning behind this idea is as follows: physical phenomena produce the technological functioning (and malfunctioning) of technological devices.[5] Therefore, designing a technological device requires knowledge of physical phenomena, and this knowledge is acquired by means of measurements and mathematization. The aim of this contribution is to outline *an epistemology of producing knowledge of physical phenomena* and to highlight, in particular, the way such knowledge of physical phenomena is produced using measurements and mathematization.

The structure of this chapter is as follows. After this introduction, the next section briefly explains what "knowledge of a physical phenomenon" must consist of in order to enable design to occur. It also addresses the presuppositions that are involved in using this knowledge and explores how knowledge of phenomena is used in scientific modeling as a crucial part of designing a technology. The question then is how such knowledge of phenomena is produced. First and foremost, how do we come to know that there *is* a phenomenon *at all?* Surely we do so by means of measurements—yet measurements produce data, not a picture of the "unobservable" phenomenon. Bogen and Woodward (1988) have proposed that phenomena are inferred from data. Their view will be discussed in section ("Data and Phenomena"). As an alternative to this view I propose that in scientific practices, predictions of the occurrence of an "unobservable" phenomenon are inferred by combining measured and observed data describing the specific physical-technological circumstances with conjectured knowledge of the phenomenon. However, this still leaves unanswered the question of how researchers come to infer to yet unknown phenomena. In section ("Formation of Scientific Concepts of Phenomena in Experimental Practice"), I follow Feest (2010) in arguing that a scientific concept of "unobservable" physical phenomena is formed by describing relevant aspects of the experimental set-up (including the experimental data). This implies that the concept of a phenomenon is inextricably entangled with aspects of the experimental set-up held responsible for its occurrence. In turn, the experimental set-up and the use of all kinds of measurement techniques enable further investigation of the phenomenon, thereby producing different kinds of data. In section ("Phenomena and Properties – their Measurement and Mathematization"), it is explained how the data thus produced in measurements are organized by means of two important epistemic strategies: by mathematization—thus generating (phenomenological) laws—and by determining quantities that are characteristic of the materials, substances, and objects involved.

Knowledge of Physical Phenomena Used for Design

In the course of explaining "how it is possible that scientific knowledge of physical phenomena enables designing," Boon (forthcoming) addresses the question "what scientific knowledge of phenomena is," focusing on those aspects of knowledge of phenomena that enable this epistemic function. This section summarizes those aspects of this account that are relevant to the topic of the current chapter.

A key part of the idea that scientific knowledge of phenomena are epistemic building-blocks for designing a technology, is recognizing that physical phenomena should not be considered as independent physical entities. Instead, physical phenomena (such as electrical conductivity, magnetic resonance or a chemical reaction) manifest at, or are produced by means of specific physical conditions and technological circumstances. This implies that a conceptual distinction is needed between "physical phenomena" and the "physical-technological environment" responsible for their occurrence.

Crucially, the design process depends on the presupposition that *given the same conditions, the same effects will occur*. This presupposition implies that, when we design something, we assume first that a phenomenon can be produced by creating (relevant aspects of) the physical-technological circumstances held responsible for its occurrence and, conversely, that given specific physical-technological circumstances the actual occurrence of specific phenomena can be predicted. Accordingly, in the engineering sciences—and, more generally, in the experimental sciences—this presupposition functions as a *regulative principle* for producing and applying scientific knowledge of physical phenomena. An important feature of this epistemology is that, in the process of design the occurrence of all kinds of "unobservable" phenomena—such as chemical reactions, electrical conduction, or transfer of compounds between different phases—is assumed based on established scientific knowledge of these phenomena, often without checking whether these phenomena actually occur.

Scientific knowledge of phenomena that can be used in the process of design, therefore, consists of more than a description of something that can be directly observed. It consists first of the scientific concept of the phenomenon. Additionally, however, it entails knowledge of physical conditions and (if relevant) the technological circumstance at which the phenomenon manifests. It also consists of mathematical equations (i.e. laws) that represent the phenomenon as a function of (some of the known) causally relevant conditions of its physical-technological environment, by means of which the quantitative effects of physical conditions and technological circumstances can be calculated.

The aim of the design process is to work out how a technological function (e.g. removing toxic compounds from an industrial waste gas (see Kenig et al. 2001)) can be constructed in terms of the physical phenomena and the physical-technological circumstances that produce this function. This usually involves the construction of scientific models that are based on knowledge of potentially relevant physical phenomena (P_1, ... P_n) and physical-techno-logical circumstances (including knowledge about their mutual interactions) by means of which a physical phenomenon (P_T) that is held to be responsible for the technological function is generated. In this brief example, the techno-logical function "waste gas cleaning" is generated by a physical phenomenon (P_T) that is called "reactive-absorption of toxic compounds in a gas into a fluid." The scientific model eventually represents how the desired physical-phenomenon (P_T) is generated in terms of all kinds of interacting phenomena (P_1, ... P_n) and physical-technological circumstances, such as different kinds of transfer and dissolution processes of toxic compounds in the gas- and liquid-phase, and different kinds of chemical reactions.

Constructing these scientific models is an inherent part of technological design. The models enable further investigation of how the technology can be built and of the technological production of the technological function. For example, scientific models make it possible to create computer programmes capable of performing simulations by means of which the technology can be investigated. They also enable the design of experimental set-ups in which contributing physical phenomena (P_1, ... P_i) can be investigated in isolation.

Data and Phenomena

What are phenomena, and how are they identified if they are not directly observable? Bogen and Woodward (1988) developed an account of phenomena that seeks to do justice to scientific practice by distinguishing between data and phenomena (see Woodward 2011). Loosely speaking, data are the observations reported by experimental scientists, while phenomena are objective, stable features of the world whose existence scientists infer on the basis of reliable data. According to Bogen and Woodward, the melting point of lead is *inferred* or *estimated* from patterns in observed data; it is not *determined* by observing the result of a single thermometer reading (see Bogen and Woodward 1988: 308). Hence, their argument for distinguishing between data and phenomena is that *data*, for the most part, can be straight-forwardly and uncontroversially observed (e.g. the thermometer readings and the observation that a solid is melting) whereas most *phenomena* are not observable. This distinction is relevant because, according to them, data, although observable, are idiosyncratic to particular experimental contexts and typically cannot occur outside of those contexts. At the same time, data play the role of evidence for the existence of phenomena:

> [D]ata are far more idiosyncratic than phenomena, and furthermore, [...] their production depends upon highly irregular coincidences involving a great number of different factors. It follows that explanations of data, when they can be given at all, will be highly complex and closely tied to the details of particular experimental arrangements. As we vary the method used to detect some phenomenon, and other details of the experimental design, the explanation we must give of the data will also vary, often in rather fundamental ways.
>
> (ibid.: 326)

I agree with Bogen and Woodward that a distinction must be made between data and phenomena. The question is, however, how are phenomena inferred from data? Bogen and Woodward discuss two possibilities: phenomena are inferred (1) from patterns of data (e.g. by means of statistical inference) or (2) by means of "inference to the best explanation." The second option implies that descriptions of phenomena are theories, an implication they seek to avoid for obvious reasons. Concerning the first option, however, the two co-authors leave open the question of how scientists infer phenomena from data. It is a question which has been debated by several authors.

One of their critics is James McAllister (see 1997; 2011). He summarizes their view as the claim that the function of scientific theories is to account for phenomena, which Bogen and Woodward describe as both investigator-independent constituents of the world and as corresponding to patterns in data sets. Yet according to McAllister this view is incoherent. He proposes instead that phenomena are investigator-relative entities. Each one of the countless patterns exhibited by data sets has an equally valid claim to the status of phenomenon: each investigator may stipulate which patterns correspond to phenomena for him or her. Below, it will become clear that I agree with McAllister on the first point. However, I also note that the epistemic uses of observed and measured data suggest that scientific researchers agree on epistemic strategies for their organziation (see section "Phenomena and Properties – their Measurement and Mathematization").

Bruce Glymour (2000) also points out that Bogen and Woodward fail to state how scientists discern or discover phenomena in the first place (see ibid.). Bogen and Woodward claim that phenomena do not explain data. But if this is so, then we are bound to ask whether phenomena are merely summaries of data. Or is there something more to phenomena than just patterns, summaries of data, or statistical features? If so, what could this be? Glymour argues that there is not. According to him, scientists infer patterns from data by means of statistical analysis. If we accept this argument, then McAllister (1997) is mistaken in thinking that the choice about "which patterns to recognize as phenomena" can only be made by the investigator on subjective grounds. Furthermore, Glymour argues that, according to Bogen and Woodward, phenomena are nothing more than summaries of data, which can be taken to imply that phenomena coincide with patterns in data. Therefore, Bogen and Woodward are mistaken in thinking that a distinction between phenomena and data is necessary. Instead, according to Glymour, talk of phenomena is superfluous. Certainly Glymour makes a powerful argument to contest Bogen and Woodward's position. However, the argument I seek to render plausible here is that a conceptual distinction between "data" and "phenomena" (as well as some other conceptual distinctions proposed in this chapter) is crucial for pragmatic reasons—namely, to facilitate epistemic uses of measured and observed data.

At this point, it should be recognized that the distinction between data and phenomena proposed by Bogen and Woodward must be understood in the context of efforts to solve two related issues in the philosophy of science: how can observations generated by means of experiments constitute evidence for theories, and how can the theory-ladenness of observation be circumvented. Against this background, Bogen and Woodward propose that facts about phenomena—rather than data—are explained by theories:

> In undertaking to explain phenomena rather than data, a scientist can avoid having to tell an enormous number of independent, highly local, and idiosyncratic causal stories involving the (often inaccessible and intractable) details of specific experimental and observational contexts. He can focus instead on what is constant and stable across different contexts.
>
> (Bogen and Woodward 1988: 326)

Rather than focus on philosophical issues concerning the justification of theories by means of measurements, my aim here is to understand how the design process is enabled by scientific knowledge of physical phenomena. As a consequence, my account of physical phenomena contradicts Bogen and Woodward on two important points. First, Bogen and Woodward seek to avoid portraying phenomena to be some kind of low level theories, whereas in my account, "unobservable" phenomena are conceptualized, a process involving both empirical and theoretical content (see Boon 2012). Second, while I agree with Bogen and Woodward's claim that physical phenomena exist independently of us, I also argue that phenomena are not independent of their physical and (where relevant) technological environment. Phenomena are not independent, "self-enclosed," "free-floating" physical entities, so to speak.

In order to account for this ontological point of view, I have proposed a conceptual distinction between physical phenomena and the physical-technological environment causally relevant to their manifestation (see Boon *forthcoming*). As a consequence, scientific knowledge of physical phenomena involves knowledge of the causal influences exerted by their physical-technological environment. Accordingly, and contrary to Bogen and Woodward, I claim that the "highly complex details of experimental arrangements producing the data" are a relevant part of knowledge of the phenomenon. Researchers need to figure out which of these physical and technological details are causally relevant to the phenomenon and which are not. This latter aspect of my account is supported by the *regulative principle* that the same physical-technological circumstances will bring about the same effects.

Accordingly, one way in which phenomena are inferred from data is based on this principle. If researchers possess scientific knowledge of phenomena $P_1, \ldots P_n$, and also know the physical-technological circumstances of a specific "data-producing experimental set-up," this knowledge enables them to infer the occurrence of physical phenomena P_i in that system, even if the system is very different from the experimental set-ups by means of which the individual phenomena P_i were discovered and/or investigated. If this account is correct then it serves to explain, contrary to Glymour (2000), why a conceptual distinction between descriptions of patterns of data and descriptions of physical phenomena is crucial for pragmatic reasons. Without such a distinction, it would be unclear how to apply knowledge (i.e. knowledge of mere data patterns gained by means of a specific experimental set-up, rather than knowledge of phenomena occurring in specific physical-technological conditions) to another system, let alone how to apply it in designing another system—for, as Bogen and Woodward put it, the data are idiosyncratic to the system that produced them, to which I would add that physical phenomena are idiosyncratic to the specific physical-technological conditions that produced them.

Formation of Scientific Concepts of Phenomena in Experimental Practice

The broader aim of this contribution is to explain "how it is possible that scientific knowledge of physical phenomena enables designing." In order to answer this question, I contend that the trick is precisely *not* to split it into two apparently obvious, separate questions: how scientific knowledge of physical phenomena is possible and, next, how it is possible that this knowledge enables design. The crux lies in recognizing that researchers involved in experimental practices produce knowledge of phenomena *in such a manner that it enables epistemic uses*. For instance, knowledge produced by means of experiments must be such that it enables new experiments to be designed and their outcomes (i.e. the physical phenomena produced by these experiments) to be predicted. In the philosophy of science, designing new experiments that are aimed at generating phenomena that are predicted by tentative knowledge hypothesized in earlier experiments is commonly interpreted as a methodology initially intended to test the hypothesis (e.g. to test whether the purported phenomenon really does exist). Yet in actual experimental practice this approach may also be interpreted differently: preliminary knowledge hypothesized in earlier experiments (e.g. a hypothesized physical phenomenon or property) can be seen as enabling the design of new experiments which in turn facilitate further investigation of the purported object of research (i.e. the phenomenon or property), thereby generating new knowledge of it—notably, this may also involve its rejection. The hypothesis that describes the purported physical phenomenon or property is a scientific concept. Uljana Feest (2008; 2010) proposes an account of scientific concepts that explains this further. She proposes that we

> [t]hink of the descriptive features of a concept not in terms of whether they can adequately represent the object under investigation, but how they enable experimental interventions in the process of investigating the purported or ill-understood object. The basic idea here is that concepts figure as tools for the investigation of such objects. As such they can contribute to experimental knowledge generation, but they can also be refined and discarded in the process.
>
> (Feest 2010: 177)

She continues:

> The basic point here is that we cannot even begin to study the purported object of research … unless we work with a preliminary understanding of how to empirically individuate the objects that possess it. Operational definitions function as tools to this end by providing paradigmatic conditions of application for the concepts in question.[6]

In brief, Feest (2010) argues that concepts of (in my case) phenomena are formed by creating operational definitions of them; these definitions are cast in terms of a description of a typical, paradigmatic experimental set-up believed to generate data that are indicative of the phenomenon specified by the concept. Furthermore, as a consequence of this account, the descriptive features of these concepts do not initially constitute an adequate representation of the phenomenon. Instead, according to Feest, concepts are tools which enable experimental intervention in the domain of study, thereby generating knowledge about the phenomenon.

If this account is correct, it implies that: (1) the actual conception of a phenomenon is enabled by the description of aspects of an experimental set-up and (2) the resulting scientific concept is entangled with that description. This account explains how it is possible that scientific knowledge of physical phenomena enables design. When designing advanced technologies, researchers do not need knowledge of phenomena independent of the physical-technological environment responsible for their occurrence or manifestation. On the contrary, they need knowledge of the physical effects produced by a physical-technological environment (e.g. as generated by means of the experimental set-up) and, more specifically, they need to know which features of this environment are crucial for the occurrence of that effect. This is exactly what an operational definition of a phenomenon such as the one proposed by Feest (2010) seems to provide. In other words, this account explains how scientific concepts of phenomena (e.g. objects, processes, properties) are formed so that these concepts can be put to epistemic use in design processes.

In Boon (2012), I elaborate on the account of scientific concepts proposed by Feest (2008; 2011), arguing that the process of inferring from the description of aspects of an experimental set-up an operational definition of a phenomenon, which, in turn can be used as a scientific concept involves subsuming this description under more abstract concepts, such as naming it as an "object," a "property," or a "causal relationship," and under theoretical concepts, such as "force," "energy," "fluid," etc. I argue that subsuming an empirical description under such abstract and theoretical concepts makes them theoretical rather than strictly empirical, as it introduces new epistemic content that expands on what is empirically known and is therefore also hypothetical. It is exactly this additional epistemic content that enables asking new questions by means of which the investigation of the phenomenon moves forward. Furthermore, the additional abstract and theoretical content enables epistemic uses of these concepts in new circumstances, as will be shown below.

Examples of phenomena—also called properties—in the engineering sciences that have been conceptualized by means of paradigmatic experiments include material properties such as "elasticity," "viscosity," "heat content," "melting point," "electrical resistance," "thermal conductivity," "magnetic permeability," "physical hysteresis," "crystallinity," "refractivity," "chemical affinity," "wavelength," "chemical diffusivity," "solubility," "electrical field strength," "super-conductivity," and "atomic force."

The concept of each of these properties is related to experiments by means of which they were initially defined. Hooke's experimental set-up, for instance, in which the extension of a spring was measured as a function of its weight, can be regarded as a paradigmatic experiment by means of which the property "elasticity" was operationally defined. The description of the paradigmatic experiment might be formulated as follows: "to measure the reversible (and proportional) extension of a spring by a weight," which is the observable phenomenon. The preliminary operational definition of "elasticity" derived from it could be rendered as "the property of a spring to reverse its stretch when extended by a weight." Accordingly, the description of the paradigmatic experiment is subsumed under a more abstract concept (e.g. the concept "property") and also—as elasticity is conceived of as a kind of force—under the theoretical concept "force," which results in the scientific concept "elasticity" being defined as "the measurable *property* of an object to reverse a deformation imposed by a *force*."

In other words, researchers infer an operational definition of a phenomenon from a description of a paradigmatic experiment: the definition is cast in terms of a description of the paradigmatic experimental set-up. In a subsequent step the operational definition, by being interpreted as a definition of a *property* and by interpreting the observed phenomenon in terms of theoretical concepts, is turned into a scientific concept which can be applied to situations that differ from the paradigmatic experimental set-up: wherever the reversible deformation of an object occurs, we attribute the property "elasticity" to the object and assume that it is a quantifiable property, independent of the kind of object, the kind of matter and the kind of force involved. Therefore, the concept "elasticity" refers to a qualitative and quantifiable *property* of materials or substances while at the same time expressing aspects of the paradigmatic experiment significant for the occurrence of elasticity.

Note that, from a theory-oriented perspective, the epistemological approach in Hooke's experiment is interpreted differently. Van Fraassen (2012), for instance, may critically ask: "*what* quantity does Hooke's measurement measure?," going on to argue that this involves a theory-dependent answer: "Whether a procedure is a measurement and, if so, what it measures are questions that have, in general, answers only relative to a theory." Van Fraassen refers to Galileo's design of an apparatus to measure the force of a vacuum (in his *Dialogues Concerning Two New Sciences*) and argues that, from Galileo's point of view, this apparatus measures the magnitude of the force of the vacuum, although from a later point of view it is measuring a parameter absent from Galileo's theory, namely, atmospheric pressure. However, in many cases, experimental findings precede theory. Furthermore, whether experimental findings are interpreted as measuring "something" also depends on aspects of the experiment itself, such as its stability and reproducibility.

Hence, although I am not in disagreement with van Fraassen (2012), one of the consequences of shifting the focus to the role of experiments in producing and investigating physical phenomena, as proposed in this contribution, is that experimental practices may also give rise to a different epistemology. The proposal made here is that the interpretation of experimental findings involves formulating a scientific concept in terms of an operational definition and subsuming this empirical description under abstract and theoretical concepts. The covering concepts, such as "property" and "force," are not initially derived from theories, as van Fraassen suggests, but have first and foremost an everyday meaning; applying them in contexts beyond their everyday uses in the ways just mentioned makes them theoretical (see also Chang 2012).

Does this account indeed provide an understanding of how researchers produce scientific knowledge of phenomena such that it enables epistemic uses in the design process? In line with Feest (2010), I suggest that the scientific concept thus formed enables additional experimental investigation of the purported phenomenon *because* it is phrased in terms of a description of a paradigmatic experimental set-up. In sum, the scientific concept together (and entangled) with knowledge of the paradigmatic experimental set-up make it possible to investigate the phenomenon or property in varying physical conditions and technological circumstances. In such experimental research, the space of causally relevant technological and physical variables is explored, wherein the original physical-technological conditions of the experimental set-up will be varied and extended using all kinds of often newly developed measurement techniques.

Phenomena and Properties—their Measurement and Mathematization

Authors in the philosophy of science, such as Bogen and Woodward, do not usually distinguish between properties and phenomena, whereas scientific practices do. It was suggested above that in the distinct uses of these terms, a phenomenon is the actual manifestation of a property and that, conversely, a property is a capacity that manifests under specific conditions. Yet scientific practices employ an additional distinction, that is, between phenomena and measurable quantities that are characteristic of a material or object (such as a technological device). Measurable quantities are also called characteristic or specific properties but are often referred to as just "properties of a material or object."[7] In this section, I seek to elucidate how the determination of characteristic quantities of materials and objects is important as an epistemic strategy for producing knowledge of physical phenomena.

Experimental investigations of a purported phenomenon, such as those outlined in the previous section, produce different kinds of large amounts of data. In order to be useful for performing epistemic functions, these data must be efficiently organized. One of the well-known strategies in scientific research for doing so is to establish mathematical relationships (e.g. proportionality) between measured data.[8] Hooke's law, for instance, describes the extension of a spring, X, as a function of the exerted force, F, and a constant factor, k, the elasticity coefficient of a spring. Stated more generally, these kinds of equations describe the phenomenon (e.g. "deformation of an elastic object by means of exerting a force") as a function of variable quantities (i.e. causally relevant technological circumstances such as length and width of the spring, and physical conditions such as temperature and pressure) and some more stable quantities that characterize the substance, material, object, or system under study (e.g. the elasticity coefficient of a material or object). Accordingly, in constructing these kinds of mathematical equations for describing measured data (i.e. phenomenological laws), a conceptual distinction is made for pragmatic reasons between (1) variable quantities typical of the phenomenon, (2) variable physical and technological quantities affecting or determining the phenomenon, and (3) more stable quantities characteristic of the substances, materials, objects, and systems involved.

Generally speaking, the aim of experimental practices is to characterize substances, materials, objects, and systems in terms of stable, quantifiable physical properties, that is, stable quantities called *characteristic properties*. These stable quantities are derived from measurements by converting measured data to a quantity *per* characteristic unit of the substance, material, object or systems, such as per unit of mass, molecules, electrons, length, surface, volume, time, or temperature. For instance, the density of a material is the measured weight of this material *per* characteristic unit of volume (e.g. cubic meter) of this material; the elasticity coefficient of a spring is its extension *per* unit of length of the spring and per unit of mass causing its extension; the heat transfer coefficient of a material is the measured Joules transferred *per* unit of time, per unit of surface, per unit of length (thickness), and per unit of temperature difference between the two surfaces of the material.

Note that the inference from measured data to characteristic stable quantities is only justified if the proportionality has been experimentally tested. Also note that the values of these stable quantities are usually still dependent on causally relevant conditions. The density and the elasticity coefficient of a specific material, for instance, are affected by its temperature. Similarly, in the case of such causal influences on "stable" quantities, researchers will deal with this using the same epistemic strategy, namely, constructing mathematical equations that describe the property (such as the elasticity coefficient) as a function of *variable quantities* (i.e. causally relevant physical conditions and technological circumstances). The latter equations may entail yet other stable quantities that characterize the substance, material, object, or system under study (e.g. its molar weight, its specific heat constant). Hence, again and again, the same epistemic strategies of experimentation and mathematization are used in producing scientific knowledge of phenomena and properties.

The values of characteristic properties of materials etc. are most reliably measured by standardized measurement methods.[9] These values are summarized in handbooks such as the classic *CRC Handbook of Chemistry and Physics.*[10] Significantly, any one kind of property can be determined of many different kinds of materials (e.g. the elasticity coefficient of different kinds of materials or the melting point of different kinds of metals and fluids). Conversely, any one kind of material (e.g. gold) allows for determining many different kinds of properties (e.g. its density, melting point, electrical conductivity coefficient, and elasticity coefficient). Besides being convenient for constructing mathematical equations to describe phenomena, the values of characteristic properties of materials etc. are also useful for comparing differences between materials (or substances, objects, and systems), which is important for design.

Similarly, specific physical properties of types of technological processes and systems can be determined using standardized measurement methods. Mathematical equations and values describing these quantities are summarized in engineering handbooks.[11]

Although the qualitative and quantitative measurements used to establish physical properties are reproducible, there is nothing "essential" about them. The point being made here is that physical quantities are reproducibly and stably *produced* by means of *contingent* technological instruments and measurement procedures, which reproducibly and stably *determine* the measurement outcomes.[12] In other words, given the regulative principle stating that *under the same physical conditions the same quantitative and qualitative effects will occur*, the manifestation of these quantities is inevitable, that is, their occurrence is produced and determined by the physical-technological system and procedure used.[13] However, this also implies that there is no point in claiming that materials have properties that are in some way essential. Conversely, as soon as a technologically produced property (such as "elasticity," "electrical resistance," and "melting point") has been conceptualized, this property can often be determined (in principle, although not always in practice) of many other materials as well. In other words, these properties are made manifest in other materials by means of new measurement techniques together with the concept of that new property.

Another consequence of the observation that many properties manifest only through the technological and physical conditions produced in an experimental set-up is that there is not an essential or limited set of physical properties. On the contrary, the number of different kinds of properties of substances, materials and systems increases with technological instrumentation and experimentation and with the theoretical interpretation of their outcomes. A sign of this increase can be witnessed in the *CRC Handbook* mentioned above, which contains new properties in every new edition: in the first edition of 1914, for example, all the measured physical properties covered some 100 pages while in the 94th edition of 2014 they covered more than 2,600 pages.

Expanding on the point just made, many material properties and phenomena result only from technological interventions and interactions, that is to say, their existence and/or their manifestation depends on specific causally relevant conditions brought about by means of the physical conditions of technological instruments and procedures. Why would researchers be interested in investigating them? We only have to skim through the *CRC Handbook of Chemistry and Physics* to begin guessing at the answer to this question. Why, for example, would they be interested in physical phenomena such as diffusion, heat transfer, and electrical conduction in different types of material? And why should they measure for different types of materials' characteristic properties such as the melting point, specific heat content, diffusion coefficient, electrical resistance coefficient and so on and so forth, other than for their technological relevance? Indeed, it can be said of many of the properties and phenomena that have been investigated that the researchers involved were not so much interested in them in order to test theories; instead, most properties and phenomena are studied out of an interest in potential technological applications.

Conclusions

Traditionally, the philosophy of science has assumed that theories are the ultimate aim of science and has therefore considered the role of experiments and measurements in discovering and testing scientific theories. In this contribution, the role of measurements and experiments has been considered in a different context, namely, in relation to the question of how it is possible that scientific knowledge of physical properties and phenomena enables designing—or, should we say, *inventing*—advanced technologies. The pragmatic approach taken to articulate an epistemology that accounts for the production of scientific knowledge through measurement and mathematization such that this knowledge enables design additionally gives rise to a novel pragmatic position on the character of scientific knowledge that is significant for the philosophy of science more generally: One of the points resulting from this analysis is that the explanation of successful uses of scientific knowledge, such as their uses in technology, seems not to be in need of the kind of justification which philosophers of science often seek to provide.

The crucial point in developing an explanation of *how it is possible that scientific knowledge of physical phenomena enables designing* is that this question should not be analyzed in terms of two separate questions, *how is scientific knowledge of physical phenomena possible?* and *how does this knowledge make design possible?* The crux lies in recognizing that researchers engaging in experimental practices produce scientific knowledge of phenomena *such that* it enables epistemic uses in epistemic activities such as designing. Further, from an epistemological perspective some aspects of the process of design appear to be very similar to the scientific methodology of deriving verifiable predictions that are tested in experiments, thus enabling the hypothesis in question to be tested and improved (i.e. the hypothetical-deductive method). However, focusing on the epistemic uses of scientific knowledge produced by experimental set-ups reveals that these epistemic uses are actually inextricably linked with measurable and observable aspects of the technical and physical world.

Acknowledgments

I would like to thank Olivier Darrigol and Nadine Courtenay for inviting me to speak at their seminar *The Metrological Backstage of Experiments* where I received valuable comments on the first version of this contribution. I also wish to thank Alfred Nordmann for his agenda-setting endeavors on this topic and the ZIF in Bielefeld for hosting the conference *Dimensions of Measurements* at which I presented the second version of this contribution. The research for this contribution has been supported by an *Aspasia* grant from the Dutch National Science Foundation (NWO).

Notes

1 It is worth noting that nowadays the design of, say, a house can also be advanced. Given that this is the case, the distinction identified is an intuitive one.
2 Note, however, that the notions of "property" and "phenomenon" are conceptually entangled and are often used interchangeably. Bogen and Woodward (1988), for instance, use the sentence "Lead melts at 327 °C" as an example of a *phenomenon* that is inferred from measurements (see ibid.). This suggests that we could equally describe the measured *property* as follows: "The melting-point of lead is

327 °C." Nevertheless, as pointed out in section "Phenomena and Properties – their Measurement and Mathematization" of this contribution, a conceptual distinction between "phenomena" and "properties" is relevant in terms of how experimental set-ups and measurement results are produced, organized, and utilized in scientific practices.

3   In this contribution, "physical" is meant in the broad sense, including chemical, biological, biochemical, electrical, mechanical, thermo-dynamic, hydro-dynamic (and so forth) properties. Furthermore, different kinds of things can have physical properties, including substances, materials, phenomena, objects, and technological systems. In this contribution, this set of meanings is abbreviated by referring to the "physical properties of materials and systems."

4   Examples of measurable characteristic or specific physical properties of *materials* include the elasticity coefficient, refraction index, viscosity coefficient, diffusion coefficients, heat conductivity, electrical conductivity or resistance coefficient, magnetic permeability, specific solubility (e.g. of salts or gases in a fluid), melting and freezing temperature, critical temperature, volumetric heat capacity, chemical affinity, reaction-rate coefficient, and dissociation constant. Similarly, specific properties of *technological devices* such as industrial chemical plants play a role in design. Examples of measurable physical properties in these systems include the specific mass-transfer coefficients (e.g. for the transfer of a compound from the gas phase to the liquid phase in a mechanically stirred fluid), the specific mixing time (e.g. of a mechanically stirred fluid), and specific heat transfer coefficients. In these latter examples, "specific" means "per unit significant to the system," i.e. per unit of time, length, volume, mass, temperature, energy input etc.

5   My account of "technological function" can be found in Weber et al. (2013: 33).

6   Ibid. Chang presents an overview of "Operationalism" in Chang (2009).

7   The two terms, "property" and "quantity" are often used interchangeably. How are they related? The *Joint Committee for Guides in Metrology* (VIM 2012) defines "quantity" as "a property of a phenomenon, body, or substance, where the property has a magnitude that can be expressed as a number and a reference" VIM (2012). "International Vocabulary of Metrology—Basic and General Concepts and Associated Terms (VIM)," Document produced by Working Group 2 of the Joint Committee for Guides in Metrology (JCGM/WG 2), at http://www.bipm.org/utils/common/documents/jcgm/JCGM_200_2012.pdf.

8   In Boon (2011), I argue that data produced in experiments can be interpreted in two different ways: causal-mechanistically and mathematically. These two perspectives produce distinct scientific results, which are connected by means of the target system (the experimental set-up), but cannot be reduced to each other. Conversely, they enable distinct kinds of epistemic uses. In the current contribution, it is argued that scientific knowledge of a phenomenon required for designing involves both types of knowledge: the scientific concept presenting a causal or causal-mechanistic description that is partially phrased in terms of the experimental set-up, and the mathematical formula describing the phenomenon as a function of relevant other physical and technical circumstances.

9   For example, test methods as have been documented and published through the *American Society for Testing and Materials, ASTM International*.

10   The website of this handbook http://www.crcpress.com/product/isbn/978146657 1143 states: "Celebrating the 100th anniversary of the *CRC Handbook of Chemistry and Physics*, the 94th edition is an update of a classic reference, mirroring the growth and direction of science for a century. The Handbook continues to be the most accessed and respected scientific reference in the science, technical, and medical communities. An authoritative resource consisting of tables of data, its usefulness spans every discipline."

11 For instance, *Perry's Chemical Engineer's Handbook* at http://accessengineering-library.com/browse/perrys-chemical-engineers-handbook-eighth-edition and *The Handbook of Chemical Engineering Calculations* at http://accessengineeringlibrary.com/browse/handbook-of-chemical-engineering-calculations-fourth-edition.

12 See also Cartwright's notion of nomological machines, which are considered as stably and reproducibly functioning experimental set-ups producing stable, repeatable patterns of data (see Cartwright 1983; 1989). For an expanded explanation of Cartwright's notion see Boon (2012).

13 Note that this situation is contingently dependent on the physical, practical, and technological possibility of constructing physical systems and procedures that act stable and reproducible. This holds for many physical-technological systems. However, from a pragmatic point of view, the situation is very different for systems studied in social sciences, and also when studying more complex physical systems such as those under study in medical or climate research. Concerning these kinds of systems, the regulative principle that "at the same conditions the same quantitative and qualitative effects will happen" may still be held true by scientific researchers in these practices. Yet, it is of much lesser use as a *guiding* principle, that is, as a principle that guides (*regulates*) scientific approaches.

References

Bogen, J. and Woodward, J. 1988. Saving the Phenomena. *The Philosophical Review* 972: 303–52.

Boon, M. (forthcoming). An Epistemology of Designing.

Boon, M. 2012. Scientific Concepts in the Engineering Sciences: Epistemic Tools for Creating and Intervening with Phenomena. In: *Scientific Concepts and Investigative Practice*, edited by U. Feest and F. Steinle, 219–43, Berlin, New York: Walter De Gruyter.

Boon, M. 2011. Two Styles of Reasoning in Scientific Practices: Experimental and Mathematical Traditions. *International Studies in the Philosophy of Science* 253: 255–78.

Cartwright, N. 1989. *Nature's Capacities and their Measurement*. Oxford: Clarendon Press, Oxford University Press.

Cartwright, N. 1983. *How the Laws of Physics Lie*. Oxford: Clarendon Press, Oxford University Press.

Chang, H. 2012. Acidity: The Persistence of the Everyday in the Scientific. *Philosophy of Science* 79:5: 690–700.

Chang, H. 2009. Operationalism. In: *The Stanford Encyclopedia of Philosophy* (Fall 2009 Edition), edited by E. N. Zalta, http://plato.stanford.edu/archives/fall2009/entries/operationalism/, accessed March 17, 2016.

Feest, U. 2011. What Exactly is Stabilized When Phenomena Are Stabilized? *Synthese* 182:1: 57–71.

Feest, U. 2010. Concepts as Tools in the Experimental Generation of Knowledge in Cognitive Neuropsychology. *Spontaneous Generations: A Journal for the History and Philosophy of Science* 4:1: 173–90.

Feest, U. 2008. Concepts as Tools in the Experimental Generation of Knowledge in Psychology. In: *Generating Experimental Knowledge*, edited by U. Feest, G. Hon, H.-J. Rheinberger, J. Schickore, and F. Steinle, 19–26, MPI-Preprint 340.

Glymour, B. 2000. Data and Phenomena: A Distinctions Reconsidered. *Erkenntnis* 52:*1*: 29–37.

Houkes, W. and Vermaas, P. E. 2010. *Technical Functions: On the Use and Design of Artefacts*. Dordrecht: Springer.

Kenig, E. Y., Schneider, R., and Górak, A. 2001. Reactive Absorption: Optimal Process Design via Optimal Modelling. *Chemical Engineering Science* 56:*2*: 343–50.

McAllister, J. W. 2011. What Do Patterns in Empirical Data Tell Us About the Structure of the World? *Synthese* 182:*1*: 73–87.

McAllister, J. W. 1997. Phenomena and Patterns in Data Sets. *Erkenntnis* 47:*2*: 217–28.

Van Fraassen, B. C. 2012. Modeling and Measurement: The Criterion of Empirical Grounding. *Philosophy of Science* 79:*5*: 773–84.

Weber, E., Reydon, T. A. C., Boon, M., Houkes, W., and Vermaas, P. E. 2013. The ICE-theory of Technical Functions. Metascience 22:*1*: 23–44.

Woodward, J. F. 2011. Data and Phenomena: a Restatement and Defense. *Synthese*, 182:*1*: 165–79.

Internet Sources:

The *Joint Committee for Guides in Metrology,* Document produced by Working Group 2 (JCGM/WG 2), at http://www.bipm.org/utils/common/documents/jcgm/JCGM_200_2012.pdf, accessed March 17, 2016.

CRC Handbook of Chemistry and Physics at http://www.crcpress.com/product/isbn/9781466571143, accessed March 17, 2016.

Perry's Chemical Engineer's Handbook at http://accessengineeringlibrary.com/browse/perrys-chemical-engineers-handbook-eighth-edition, accessed March 17, 2016.

The Handbook of Chemical Engineering Calculations at http://accessengineeringlibrary.com/browse/handbook-of-chemical-engineering-calculations-fourth-edition, accessed March 17, 2016.

VIM (2012). International Vocabulary of Metrology—Basic and General Concepts and Associated Terms (VIM). Document produced by Working Group 2 of the Joint Committee for Guides in Metrology (JCGM/WG 2), http://www.bipm.org/utils/common/documents/jcgm/JCGM_200_2012.pdf, accessed March 17, 2016.

16 Uncertainty and Modeling in Seismology

Teru Miyake

Introduction

> The nether regions of the Earth are inaccessible in the ordinary sense. Before the time of Newton, when evidence about them was nearly totally lacking, it was not necessarily unreasonable to describe the Earth in terms of models involving say a Hell, or a subterranean monster shaking itself to cause earthquakes. The subsequent growth of evidence has lowered the plausibility of such models.
>
> K. E. Bullen (1975)

Seismology is the science of inferring properties of the interior of the earth from the observation of seismic waves at its surface. Uncertainty has always been a central concern to seismologists, and it is not hard to see why. The interior of the earth is vast and could potentially contain very complicated structures, while the only observations that have so far been available have been made at the surface of the earth. Although the passage above by the seismologist Keith Bullen is stated with tongue in cheek, it expresses a real worry that seismologists have grappled with: Could our models of the interior of the earth be not just somewhat inaccurate, but radically wrong?

As I will discuss below, the abstract objects that are called *earth models* in seismology are sets (sometimes infinitely large) of parameters that are taken to represent the properties of an idealized earth. The determination of an earth model thus consists in the assignment of numerical values to each of these parameters, based on observations that have been made at the surface. The earth model is supposed to give us information about the real earth, although it is open to question whether we should think of the earth model as straightforwardly representing the real earth. If we take measurement to be the "assignment of numerals to a property of objects or events—*measurand*—according to a rule with the aim of generating reliable information about those objects or events" (Boumans 2007: 3), then the determination of earth models in seismology can be viewed as measurement, or at least similar to measurement in ways that are epistemologically relevant. There is a recent strand of the philosophical literature on measurement that takes measurement to involve the use of models in an essential way (see Boumans 2007; Tal 2012), and I believe that this literature can be brought to bear on understanding the epistemology of seismology.

The aim of this contribution is twofold. The first is to provide a brief description of the sources of uncertainty in the determination of earth models. I will describe two such sources in particular. One source comes about due to mathematical limits on what can be known about a system with practically infinite degrees of freedom, based on a finite number of observations. The other source comes about because of the lack of antecedent knowledge about both the earthquake source and the structure of the interior of the earth. The second aim of this contribution is to bring some considerations from the recent literature on measurement to bear on problems in the epistemology of seismology. In particular, I believe that the view of models as measuring instruments (see Boumans 2007) and the related idea of model-based measurement (see Tal 2012) is helpful in understanding the epistemological problems associated with the determination of earth models.

A Brief History of Earth Models

I will first provide a very brief history of earth models.[1] An earthquake happens when a slippage occurs somewhere along a fault. A sufficiently powerful earthquake will generate seismic waves, some of which travel through the interior of the earth. These seismic waves can be detected at seismographic receiving stations all over the globe. Two types of seismic waves are generated—P-waves, which are essentially longitudinal waves, and S-waves, which are essentially transverse waves. The speeds of these two types of waves differ, and depend upon the mechanical properties of the medium that the waves are passing through, such as the density and elastic moduli.

The interior of the earth is known to be layered—there are places where there are abrupt changes in the mechanical properties of the medium. Just as light waves are reflected and refracted at interfaces between two media with differing indices of refraction, seismic waves are also reflected and refracted at discontinuities between layers inside the earth. Within layers, the mechanical properties tend to vary continuously—this makes seismic waves travel along curved paths, where the specific path taken will depend upon the properties of the medium that the wave has traveled through. Some waves go directly from the seismic source to the receiving station, while other waves reach the receiving station after having reflected off the surface of the earth or having been reflected or refracted at discontinuities. Since these various waves travel along different paths, they are detected at the receiver at different times after the seismic event. Different such wave arrivals are called *phases*.

If we assume that the internal structure of the earth is spherically symmetric, the times for specific seismic phases to travel specific distances will be constant. These global travel times were compiled in the form of tables in the

1930s and 40s by two teams—on the one hand, Harold Jeffreys[2] and Keith Bullen at Cambridge, and, on the other, Beno Gutenberg and Charles Richter at Caltech. From these tables, the speed of P- and S-waves at various depths within the earth can be estimated. Since the speeds of P- and S-waves depend upon the mechanical properties of the medium through which the waves pass, this can yield information about the internal structure of the earth, and the first sophisticated earth models were developed by these two groups.

In the 1960s, it became possible to bring a new type of seismological observation to bear on the problem of determining the internal structure of the earth—observations of the normal modes, or free oscillations, of the earth. These normal modes are analogous with the normal modes of a string. The motion of an idealized mathematical string with tension (τ) and density (ρ) can be approximated by a wave equation, and with the addition of certain boundary conditions (such as assuming that the ends of the string are fixed), only certain functions are possible solutions to this wave equation. These functions are called *eigenfunctions*, or *normal modes*, and the frequencies corresponding to the normal modes are called *eigenfrequencies*, or *normal mode frequencies*. Arbitrarily shaped waves traveling along the string can be represented as linear combinations of these normal modes using the theory of Fourier analysis.

Essentially the same thing can be done for the vibrations of an idealized three-dimensional object. Suppose we idealize the earth as a spherically symmetric, isotropic, perfectly elastic, nonrotating body. The free vibrations of such a body are a linear combination of the three-dimensional equivalent of the normal modes of the string—in this case, they are complicated three-dimensional functions involving spherical harmonics. For an idealized earth that is spherically symmetric, isotropic, perfectly elastic, and nonrotating, and given its rigidity, incompressibility, and density as functions of radius, the frequencies of its normal modes can be calculated. This problem is in principle solvable uniquely and exactly, although it may require integration by computational techniques. The problem that seismologists are interested in, however, is the *inverse* of this problem. By making observations of the normal modes of the earth, seismologists are trying to make inferences about its interior.

Seismologists refer to both the determination of the properties of the interior of the earth from travel times, and their determination from normal mode frequencies, as *inversion*, or *inverse problems*.[3] To make the discussion in the rest of this contribution clear, it is worth spelling out exactly what these inverse problems involve. We start with an earth model—an idealized representation of the earth. This earth model is taken to be fully characterizable by a set of parameters that represent the interior properties of the earth. Given an earth model, and given the values of its parameters, we can calculate the travel times or the normal mode frequencies we can expect to see for such an idealized earth. This is called the forward problem. The inverse problem is the problem of determining the values of the parameters of an earth model, given: (a) an earth model, and (b) a forward problem, i.e., a method for determining the observations (such as normal mode frequencies) you would expect to see for any particular assignment of values to the parameters of the earth model, and (c) a set of observations (such as normal mode frequencies).

Before we go on, I want to make a couple of terminological points about earth models. First, it is worth noting that the term *earth model* might be somewhat confusing for those who are used to the philosophical literature on models (e.g. Morgan and Morrison 1999), since the term *model*, in the way it is used in the philosophical literature, corresponds more to the method for determining the observations given any particular assignment of values to the parameters of the earth model (that is, the forward problem) than the thing that is called an *earth model* by seismologists. Second, there are two different ways in which one might individuate earth models. On the one hand, we might count one earth model as different from another only if they parameterize the interior of the earth in different ways. On the other hand, we might count one model as different from another even in cases where the two models are parameterized in the same way, but if they assign different sets of values to the parameters. In agreement with normal usage in seismology, when I write "different earth models," the latter usage is understood. In the case of the former usage, I will write "different parameterizations."

Uncertainty Arising due to Mathematical Limits

An important issue for seismologists has been the question of to what degree these inverse problems are solvable. Now, of course, a worry that one might have is that there will be errors on the observations, so that a unique earth model cannot be determined due to observational error. This is a legitimate worry, but it turns out that seismologists were also concerned about another nonuniqueness problem—one having to do with the limits of what can be known about a vastly complicated system, given only a finite number of observations.

First, let us consider how one might go about solving an inverse problem. Given the characterization of inverse problems above, many philosophers would, I think, immediately think of the hypothetico-deductive method. The problem is to determine the values of the parameters of an earth model, given a method for calculating the observations you would expect to see for that model, and a set of actual observations. So one way of solving this inverse problem is as follows: you assign a certain set of values to the parameters of an earth model, then calculate the observations you would expect to see for that model. Then you compare these expected observations with the actual observations. If the expected and actual observations disagree, you would change the parameter values and repeat the comparison. If you find a set of parameter values for which the expected and the actual observations agree (perhaps to some degree of accuracy), you take that set of parameter values to hold for the actual earth, at least approximately.

There is a deep problem here, however, which is well known to philosophers: the possibility of empirical equivalence. There might be radically different earth models that make exactly the same predictions. If so, then a match between the predictions of a particular earth model and actual observations is no guarantee that the model is an accurate representation of the interior of the earth, for there could be a radically different but as yet untested model that makes exactly the same predictions. This is, of course, the classic problem of underdetermination.

This was, in fact, a real worry among seismologists in the 1960s. In 1968, the geophysicist Frank Press published a paper in which the aim was to explore the possibility that the assumptions made in the production of earth models at the time were wrong, and that there could be radically different earth models that have the same degree of fit to the observations (see Press 1968). Press used a computer to produce 5 million different earth models, calculated the moment of inertia, travel times, and normal mode frequencies for each model, and compared these values against observed data. Among the 5 million models, Press found six that agreed with observations but rejected three of them as being implausible. But the three models that he found to be plausible were significantly different from each other, especially in the upper mantle.

Ultimately, the problem rests with the fact that the interior of the earth has practically infinite degrees of freedom, whereas the number of observations of the earth is finite. For example, if you idealize the interior of the earth as a continuous medium that is spherically symmetric, homogeneous, isotropic, and linearly elastic, then you can completely characterize the mechanical properties at each point in the earth using three parameters—density, and two elastic moduli. The assumption of spherical symmetry then allows us to completely characterize the interior of the earth by assigning a value for these three parameters at each distance r from the center of the earth—that is, by three functions that represent the density and the two elastic moduli as functions of r. Under these idealizations, then, an earth model is just a set of three functions of r.

Suppose we wanted to completely determine such an earth model based on observations of seismic waves at the surface of the earth. Consider for now just the function $\rho(r)$ that represents the density of the interior of the earth as a function of radius. Let a be the radius of the earth. Then $\rho(0)$ would be the density at the center of the earth, and $\rho(a)$ would be the density at the earth's surface. If we take $\rho(r)$ to be an arbitrary function of radius, it can take any value at any point on the interval $[0, a]$. Of course, we can put some constraints on it. For example, we know that there are several layers within the earth with abrupt changes in the mechanical properties in between. Within each layer, we might surmise that $\rho(r)$ will be fairly smooth. Still, $\rho(r)$ is bound to be a function with a very large number of degrees of freedom. Or, to be more precise, the density variation within the earth has a very large number of degrees of freedom, and any idealized function $\rho(r)$ that is to capture the density variation to a high degree of accuracy will also have to have a large number of degrees of freedom. But at any given time, seismologists only have a finite number of observations from which the function $\rho(r)$ must be determined. So we might wonder: to what degree can that function be determined uniquely, given a limited number of observations?

The geophysicists George Backus and Freeman Gilbert, motivated in part by the results of Frank Press, wrote a seminal series of papers (see Backus and Gilbert 1967; 1968; 1970) that explicitly considers this question. Backus and Gilbert frame their problem as an epistemological one, in which they discuss what one ought to do in the face of certain epistemic limitations—namely, the paucity of observational data relative to the number of degrees of freedom of the internal structure of the earth. One of the great contributions of Backus and Gilbert was that they found a way of taking this epistemological problem and reducing it to a precise mathematical problem. First, consider the function $\rho(r)$ on the interval $[0, a]$. We can think of a function as being an ordered set of an infinite number of values, one for each point on the interval $[0, a]$. Now, for any ordered set of n numbers, we can represent that set as a vector in an n-dimensional space (for example, an ordered triple can be represented as a vector in a three-dimensional space). So the function $\rho(r)$ can be represented as a vector in an infinite-dimensional space, a space that represents all possible ways for the density distribution in a spherically symmetric earth to be. In the literature after Backus and Gilbert, this space is often referred to as a "model space" because each vector in this space represents one possible earth model.

Now suppose that there is a function g_i for each normal mode i that maps each vector in the space of possible earth models onto a number that represents the value of the frequency of that normal mode. Suppose there are 1,000 normal modes that have been observed. Then there would be 1,000 such functions g_i. Given these functions, we can calculate, for any given earth model, the values of the frequencies of all of the normal modes. Imagine, for now, that there is one vector $\mathbf{m}_E$ in the model space that represents the *real earth*—or the closest one to the real earth, at any rate. Then, the question of to what degree the function $\rho(r)$ can be determined reduces to the question of to what degree the vector $\mathbf{m}_E$ can be pinpointed in the model space, given the functions g_i and the actually observed values of the normal mode frequencies. If you make certain assumptions about what the functions g_i are like (they must, for example, be linearized) then you can use the theory of linear vector spaces to answer this question. Backus and Gilbert show that, since the model space is infinite-dimensional, and the number of observations is finite, the vector $\mathbf{m}_E$ will be indistinguishable from an infinite number of other models.[4]

This is a somewhat troubling result, but not entirely unexpected, because according to the way in which the earth model is defined, it has infinite degrees of freedom, whereas there are only a finite number of observations. But we should also note that a huge number, in fact an infinite number, of these models will differ from each other only on the scale of, say, millimeters. Such models are effectively identical from a geophysical standpoint. So, in fact, if it can be shown that the models which agree with observation are similar enough, then the nonuniqueness problem can effectively be solved. So an important question is whether the models that agree with observation can be considered to be all the same from a geophysical standpoint, or whether there are in fact models that agree with observations, but which are different in geophysically significant ways. Backus and Gilbert develop a method for determining how different these models can be from each other and still agree with observations, now called the Backus-Gilbert resolution method. Under certain conditions, the nonuniqueness can be dealt with—if, for example, the *real earth* is in the space of possible models, and the functions g_i meet certain conditions (e.g., they must be linear functions), then the Backus-Gilbert method can be used to infer facts about the real earth.

Some readers might have noticed, however, that, in fact, these conditions are never met exactly. For example, the *real earth* cannot be in the space of possible earth models, at least under the idealization given above—each model in the space is spherically symmetric, homogeneous, isotropic, and linearly elastic, but the real earth is none of these. Now, of course, from a practical standpoint, we might still be fairly confident that we can use these methods to extract information about the real earth. But from an epistemological standpoint, we would want to further understand what effect these idealizing assumptions have on our claims to having knowledge about the interior of the earth. I will thus discuss these assumptions in the next section.

Uncertainty Arising due to Idealizing Assumptions

First, let us consider again how the problem of inferring properties of the interior of the earth from observations of travel times and normal mode frequencies is solved. In effect, what is done is that the earth is first parameterized in a particular way—note again that there could be a finite number of parameters, or there could be an infinite number. This parameterization defines the space of possible earth models. There is a forward problem, that is, a way of calculating for any given earth model what the expected observations are. And there is a set of actual observations. Now, suppose we are looking for a vector in the space of possible earth models that represents the *real earth*. Then, under certain conditions, we can use mathematical techniques to find the set of all earth models that agree with observations, and we can further examine the properties of the earth models in this set in such a way that we can extract information about the real earth.

There are several problems, however. As I mentioned above, it could be that none of the models in the space of possible earth models plausibly represents the actual earth. For example, take the assumption of isotropy. It is known that at a fine enough grain, the interior of the earth is not actually isotropic, and there very well could be parts of the earth that have significant departures from isotropy. But if the inverse problem is set up in such a way that isotropy is built into the earth model from the start, then the solving of such an inverse problem will never show us what parts of the earth are anisotropic. And apart from these assumptions that are built right into the parameterization of the earth models, there are also assumptions that are made in the calculations of normal mode frequencies that are not straight-forwardly reflected in the way the model space is defined. For example, when the normal mode frequencies are calculated, it is assumed that when waves propagate through the medium, there is no attenuation of the waves due to the anelasticity of the medium.

Idealizations are not a problem if we know in advance that the idealizations are "safe" to make. This is often the case with systems that are *antecedently familiar* to us. For example, suppose I wanted to calculate the distance a baseball would fly if projected at a certain angle at a certain velocity. When I make this calculation, I can safely ignore factors like air resistance because I am familiar with objects like baseballs, and know that air resistance will not make a big enough difference to the calculation. Of course, if I needed to make an accurate enough calculation, I would have to take it into account. Similarly, if there, say, happened to be a layer within the earth that is signifi-cantly anisotropic, and this made a significant difference to the observations seismologists have, then they would want to take this layer into account. But, in general, seismologists do not know antecedently whether such layers exist, or where they might be.

Moreover, if such a significantly anisotropic layer of the earth existed, it is somewhat of a puzzle how solving an inverse problem would show us that such a layer exists, since assumptions like isotropy are typically built into the model space from the start. It seems, then, like there is a deep problem: inverse problems work by postulating a space of possible models and then selecting out an observationally acceptable subset of these models by comparison with observations. In order to enable these methods, the models must be idealized due to considerations of mathematical tractability. But it is not known in advance, in this case, exactly which idealizations are "safe" to make.

How, then, do you go about figuring out which features you need to incorporate into the models? There are at least a couple of ways. First, one way would be to examine deviations between what your current model says the observations should look like, and the actual observations. If there are systematic deviations in these residuals, this will tell you that there is some property or feature that is not being taken into account. An analysis of these residuals may yield clues about specifically what that property or feature is. Another thing seismologists might do is to compare two different models with different sets of assumptions, and examine the deviations between these models carefully. This might yield clues towards things that are being left out of these models. For example, in a review article, the seismologists Dziewonski and Woodhouse (1987) discuss a case where an examination of the differences between two different isotropic earth models led to the discovery that the inner core of the earth is anisotropic (see ibid.).

I call this process of trying to find previously unknown but significant features of the earth *exploration*. I thus think that one of the important uses of models in seismology is in this process of exploration, and we should understand the recent history of seismology in terms of the building of models and the use of these models for exploration.

Measurement and Exploration: Some Considerations from the Model-Based Account of Measurement

What makes seismology particularly difficult is that the aim is to extract knowledge about a vast, complicated object, but the ways in which we can observe this object are limited in certain ways. We cannot isolate this system in a laboratory in order to shut out any extraneous causal influences. We cannot vary parameters of this system at will. We will probably never be able to get up close to most parts of this system, in the deep interior of the earth, so we will never be able to bring measuring instruments down there. So seismologists must infer properties of the interior of the earth through the solving of inverse problems as I have described above.

Now, the epistemology of seismology has some perhaps surprising parallels to the epistemology of economics. The aim of economics, as well, is to extract knowledge about a vast, complicated system for which our methods of observation are limited in certain ways. In economics, as well, we cannot build physical instruments to measure properties of economic systems, so economists must build models, and use these as measuring instruments (see Boumans 2007). More recently, Tal (2012) has argued that models are an essential part of all measuring practices.

I have already suggested that we should take the determination of earth models through the solving of inverse problems to be a kind of measurement, and this fits the model-based account of measurement (see Tal 2012). There are some differences. Perhaps the biggest difference is that in the solving of inverse problems, seismologists sometimes attempt to determine not just a finite number of parameters, but functions, which can be thought of as an infinite number of parameters. And because of this, the nonuniqueness problem that I discussed above can arise. Nevertheless, I think that the determination of earth models can be regarded as at least analogous to model-based measurement on Tal's account, and they are similar in epistemologically relevant respects.

Here, let me outline one way in which some ideas from the model-based account of measurement might illuminate the epistemology of seismology. The account given here should only be taken to be some preliminary thoughts about how the model-based account might be applied to seismology. I mentioned above that there can be uncertainty in the determination of properties of the interior of the earth through the solving of an inverse problem, because an inverse problem necessarily involves an idealized model, and there could be significant differences between the model and the actual earth. What is interesting here are some comments that Tal makes about *nontrivial* measurement procedures—"ones that involve new kinds of instruments, novel operating conditions or higher accuracy levels than previously achieved for a given quantity" (Tal 2012: 65). These are cases where you are trying to develop new measuring techniques in order to extend the scope of measurability. Often, you need to develop new measuring procedures that are radically different from accepted procedures, and the natural question that arises is: "Are we actually measuring the same quantity?" Tal comments as follows:

> Historically, the process of extension has almost always been conservative. Scientists engaged in cutting-edge measurement projects usually start off by dogmatically *supposing* that their instruments will measure a given quantity in a new regime of accuracy or operating conditions. This conservative approach is extremely fruitful as it leads to the discovery of new systematic errors and to novel attempts to explain such errors. But such dogmatic supposition should not be confused with empirical knowledge, because novel measurements may lead to the discovery of new laws and to the postulation of quantities that are different from those initially supposed. Instead, this sort of dogmatic supposition can be regarded as a manifestation of a regulative ideal, an ideal that strives to keep the number of quantity concepts small and underlying theories simple.
>
> (ibid.: 66)

The idea is this. When you start developing new techniques for measuring quantities, you will usually have discrepancies between these new techniques and old ones. But you do not simply assume that they are measuring different quantities. You assume that they are measuring the same quantity, and try to figure out why these discrepancies arise. More specifically, scientists try to identify systematic errors and figure out what is giving rise to them. And, in fact, this process can lead to the discovery of new laws or factors that had not been noticed before. So convergent measurement turns out, on this view, to be a regulative ideal—one that is used to create more knowledge.

I take this idea to be rather speculative, and it is not an essential part of Tal's model-based account. But I am focusing on it here because it is analogous to certain issues that I have been considering in the epistemology of seismology. When we attempt to measure a quantity, we are presupposing a causal relationship between the quantity we are trying to measure and the position of an indicator. This causal relationship is captured in what Tal calls a "model of the measurement procedure." Normally, we try to set up measurements in such a way that the position of the indicator reliably varies in a known way with any variation in the quantity we are trying to measure. Sometimes, though, we might not have control over the causal relationship between the quantity we are trying to measure, and the position of an indicator. Or, we might not have confidence that the causal relationship is known in its entirety. If we take the seismograph to be an *indicator*, which we want to use to measure certain quantities pertaining to the properties of parts of the deep interior of the earth, then we can think of seismology as an example of this sort of situation.

If we do not know whether we have accounted for the causal relationship between the quantity we are trying to measure, and the position of an indicator in its entirety, then making inferences from the indicator would be rather risky. If I want to use an ammeter to measure the electric current through a wire, but there is some influence on the ammeter needle that I do not know about, my measurement is probably going to be incorrect. Likewise, if I want to use a seismograph to measure properties of the earth's deep interior, any significant causal influences on the seismograph that I am unaware of would be problematic. So the question is: when can I be confident that my measurements are at least approximately correct?

One thing I mentioned above is that you might try to address this problem through the idea of convergent measurement—by comparison of two different models using different assumptions. And I mentioned, of course, that a problem is that two different models are always going to have discrepancies. Sometimes, those discrepancies are insignificant. But other times, those discrepancies reflect substantive differences in what the models are telling us about the interior of the earth. How do you tell one from the other?

Here is where we might think of using some ideas from the model-based account of measurement. According to what we just saw, it is the model of the measurement procedure that is crucial in solving this kind of problem. In the case of seismology, the model of the measurement procedure is the way in which the theory of waves in continuous media is used to extract information from seismograms, as well as any corrections to the seismograms, such as corrections for asphericity, anelasticity, anisotropy, the positions of any discontinuities, and so on. Seismological measurement can be considered to be what Tal calls *nontrivial* measurement—it is a case where seismologists are trying to make measurements in a regime where there is still no standard way of making the measurement. And just as with newly developed techniques for measuring temperature, say, there will be systematic errors that arise, for which seismologists have not yet accounted. For example, we do not know in advance exactly where the discontinuities are, or whether there are significant departures from isotropy or elasticity—these are things that had to be found out by seismologists. As Tal points out, in such cases, scientists sometimes make the dogmatic supposition that two different measurement procedures are measuring the same quantity—and any discrepancies are taken to be due to systematic errors that are unaccounted for. Tal claims that this can lead to the discovery of new laws and the postulation of new quantities different from those initially supposed.

Now, can something similar be said for seismology—in particular, that convergence between models can be thought of as a kind of regulative ideal?

I think yes, and it changes the way in which we think about the epistemology of seismology. Suppose there are two different ways of measuring properties of the deep interior of the earth: one, say, involving travel times of seismic waves, and another involving frequencies of the normal modes of the earth. One way to think about these models and what they say about what we know about the interior of the earth is to think that, to the extent that these models agree with each other, this can be taken to be evidence that they are, in some sense, correct. But here is another way to think about these models. There can also be discrepancies between the models. You might take these discrepancies to be indications of systematic errors that are not accounted for in these models. This can lead to the discovery of new features of the earth that have not been accounted for—that is, it can be a method of exploration. For example, as I mentioned above, an examination of the differences between two different isotropic earth models might lead to the discovery that the inner core of the earth is anisotropic. Thus, we can view earth models and other seismological models as not simply representations of the earth, but as instruments—instruments for measurement, and instruments for exploration.

Conclusion

I have discussed some problems of uncertainty that arise in seismology, one arising due to mathematical limitations in the determination of a model with an extremely high number of degrees of freedom based on a finite number of observations, and another arising due to the possibility that significant features of the earth are not being taken account of in setting up inverse problems. I take the determination of earth models to be a kind of measurement involving the use of models. The recent philosophical literature on models and measurement is thus relevant to the epistemology of seismology. I have suggested, speculatively, that certain ideas concerning nontrivial measurement procedures might be applied to seismological measurement. It remains to be seen, however, whether the application of ideas from the recent philosophical literature on models and measurement to problems of seismology will be fruitful, since there certainly are significant differences between the examples of measurement that this literature considers, and measurement in seismology.

Notes

1 Some of the expositions of seismological problems overlap with expositions in Miyake (2011; 2013). Much of the history here is based on Bullen (1975).
2 As we shall see, inferring properties of the interior of the earth from travel times is no easy feat, and it is no coincidence that Jeffreys was deeply concerned with inference and probability. I will not examine the work of Jeffreys here, however.
3 Inverse problems show up in other fields, such as computer imaging, but I believe seismologists were the first to refer to this kind of inference as inversion.
4 For a more detailed examination of this method see Miyake (2011).

References

Backus, G. and Gilbert, F. 1970. Uniqueness in the Inversion of Inaccurate Gross Earth Data. *Philosophical Transactions of the Royal Society of London, Series A, Mathematical and Physical Sciences* 266:*1173*: 123–92.

Backus, G. and Gilbert, F. 1968. The Resolving Power of Gross Earth Data. *Geophysical Journal of the Royal Astronomical Society* 16: 169–205.

Backus, G. and Gilbert, F. 1967. Numerical Applications of a Formalism for Geophysical Inverse Problems. *Geophysical Journal of the Royal Astronomical Society* 13: 247–76.

Boumans, M. (ed.) 2007. *Measurement in Economics: A Handbook.* Amsterdam: Elsevier.

Bullen, K. 1975. *The Earth's Density.* London: Halsted Press.

Dziewonski, A. and Woodhouse, J. 1987. Global Images of the Earth's Interior. *Science* 236:*4797*: 37–48.

Miyake, T. 2013. *Dealing with Underdetermination: Inverse Problems, Eliminative Induction, and the Epistemology of Seismology.* Unpublished manuscript.

Miyake, T. 2011. *Underdetermination and Indirect Measurement.* PhD dissertation, Stanford University.

Morgan, M. and Morrison, M. (eds.) 1999. *Models as Mediators: Perspectives on Natural and Social Science.* Cambridge: Cambridge University Press.

Press, F. 1968. Density Distribution in Earth. *Science* 160:*3833*: 1218–21.

Tal, E. 2012. *The Epistemology of Measurement: A Model-Based Account.* PhD dissertation, University of Toronto.

17 A Model-Based Epistemology of Measurement

Eran Tal

Introduction

Lord Kelvin famously stated that "when you can measure what you are speaking about and express it in numbers you know something about it; but when you cannot measure it ... your knowledge is of a meagre and unsatisfactory kind" (Thomson 1891: 80). Today, in an age when thermometers and ammeters produce stable measurement outcomes on familiar scales, Kelvin's remark may seem superfluous. How else could one gain reliable knowledge of temperature and electric current if not through measurement? But the quantities called "temperature" and "current" as well as the instruments that measure them have long histories during which it was far from clear what was being measured and how—histories in which Kelvin himself played important roles (see Chang 2004: 173–86; Gooday 2004: 2–9).

These early struggles to find principled relations between the indications of material instruments and values of abstract quantities illustrate the dual nature of measurement. On the one hand, measurement involves the design, execution, and observation of a concrete physical process.[1] On the other hand, the outcome of a measurement is a knowledge claim formulated in terms of an abstract and universal concept—e.g. mass, current, temperature or duration. How, and under what conditions, are such knowledge claims warranted on the basis of material operations?

Answering this last question is crucial for understanding how measurement produces knowledge. And yet until very recently the philosophy of measurement offered little by way of an answer. The philosophical literature on measurement has traditionally focused on either the mathematical structure of measurement scales (see Krantz et al. 1971) or the metaphysics of quantity (see Swoyer 1987: 235–90; Michell 1994: 389–406). Epistemological concerns about measurement were discussed by Pierre Duhem (Duhem 1962 [1906]) and became briefly popular in the 1920s (Campbell 1920; Bridgman 1927; Reichenbach 1958 [1927]) and again in the 1960s (Carnap 1995 [1966]; Ellis 1966), but have otherwise remained in the background of philosophical discussion. It is only in the last decade that a new wave of philosophical writing about the epistemology of measurement has appeared[2] that tackles questions such as: how can one tell whether an instrument measures the quantity it is intended to? How do calibration procedures establish accuracy? Do standardization activities produce new knowledge about the quantity being standardized? How is the evaluation of measurement error and measurement uncertainty possible?

Partly drawing on this recent literature, this chapter will outline a novel systematic account of the ways measurement produces knowledge. I call my account "model-based" because it tackles epistemological challenges by appealing to abstract and idealized models of measurement processes. My examples of measurement will be drawn from metrology, officially defined as "the science of measurement and its application."[3] Metrologists are the physicists and engineers who design and standardize measuring instruments for use in scientific and commercial applications, and often work at standardization bureaus or specially accredited laboratories. This chapter will motivate the need for an epistemology of measurement, identify some of its main challenges, and show that a model-based approach holds particular promise in this field. Specifically, I will argue that metrological calibration is a special sort of modeling activity, and that this model-based conception of calibration successfully addresses certain difficulties associated with the philosophical study of measurement.

The Epistemology of Measurement

The epistemology of measurement is a subfield of philosophy concerned with the relationships between measurement and knowledge. Central topics that fall under its purview are the conditions under which measurement produces knowledge; the content, scope, justification, and limits of such knowledge; the reasons why particular methodologies of measurement and standardization succeed or fail in supporting particular knowledge claims; and the relationships between measurement and other knowledge-producing activities such as observation, theorizing, experimentation, modeling, and calculation (see also Tal 2016). The pursuit of research into these topics is motivated not only by the need to clarify the epistemic functions of measurement, but also by the prospects of contributing to other areas of philosophical discussion concerning reliability, evidence, causality, objectivity, representation, and information, among other topics.

As measurement is not exclusively a scientific activity—it plays vital roles in engineering, commerce, public policy, and everyday life—the epistemology of measurement is not simply a specialized branch of philosophy of science. Instead, the epistemology of measurement is a subfield of philosophy that draws on the tools and concepts of traditional epistemology, philosophy of science, philosophy of language, philosophy of technology, and philosophy of mind, among other subfields. It is also a multidisciplinary subfield, engaging with measurement techniques from a variety of disciplines, as well as with the histories and sociologies of those disciplines.

Indications vs. Outcomes

As already noted, one of the central aims of the epistemology of measurement is to elucidate the relationships between the concrete and abstract aspects of measuring, and particularly between *instrument indications* and *measurement outcomes*. An indication is a property of a measuring instrument in its final state after the measurement process is complete. Examples of indications are the numerals appearing on the display of a digital clock, the position of an ammeter pointer relative to a dial, and the pattern of diffraction produced in X-ray crystallography.

Note that in the current context the term "indication" carries no normative connotation. It does not presuppose reliability or success in indicating anything, but only an *intention* to use such outputs for reliable indication of some aspect of an object of interest. Note also that indications are not numbers: they may be symbols, visual patterns, acoustic signals, relative spatial or temporal positions, or any other sort of instrument output. Nonetheless, indications are often represented by mapping them onto numbers, e.g. the number of "ticks" the clock generated at a given period, the angle of displacement of the pointer, and the spatial density of diffraction fringes. These numbers, which may be called "quantified indications," are convenient representations of indications in mathematical form.[4] A quantified indication is not yet a claim about any aspect of the object or event intended to be measured, but only a mathematical description of the final state of the measuring apparatus.

A measurement outcome, by contrast, is a knowledge claim associating one or more parameter values with the object or event being measured, a claim that is inferred from one or more indications along with relevant background knowledge. Outcomes are typically expressed in terms of a particular unit on a particular scale and include, either implicitly or explicitly, a margin of uncertainty. Corresponding examples of measurement outcomes are the claim that the duration of an event measured by the clock is $x \pm U_x$ seconds, the claim that the intensity of electric current in the wire is $y \pm U_y$ amperes, and the claim that the distance between crystal layers is $z \pm U_z$ nanometers, with U being an uncertainty term.[5] Although the examples in this contribution will all be quantitative, it is important to note that measurement outcomes need not be quantitative, and may pertain to nominal or ordinal parameters.

To attain the status of a measurement outcome, a knowledge claim must be abstracted away from its concrete method of production and pertain to some quantity objectively, namely, be attributable to the measured *object* rather than to the idiosyncrasies of the measuring instrument, environment, and human operators. Consider the ammeter: the outcome of measuring with an ammeter is a claim about the intensity of electric current running through the input wire. The position of the ammeter pointer relative to the dial is a property of the ammeter rather than the wire, and is therefore not a candidate for the content of the relevant measurement outcome.[6] This is the case whether or not the position of the pointer is represented on a numerical scale. It is only once certain theoretical and statistical assumptions are made and tested about the behavior of the ammeter and its relationships with the wire and environment that one can *infer* values of electric current from the displacement of the pointer.

Quantified indications are easily confused with measurement outcomes because many instruments are intentionally designed to conceal their difference. Direct-reading instruments, e.g. household mercury thermometers, are designed so that the numeral that appears on their display already represents the best estimate of the quantity of interest on a familiar scale.[7] The complex inferences involved in arriving at a measurement outcome from an indication are "black-boxed" into such instruments, making it unnecessary for users to infer the outcome themselves. Regardless of whether or not users are aware of them, such inferences form an essential part of measuring. They link claims such as "the pointer is between the 0.40 and 0.41 marks on the dial" to claims like "the current in the wire is 0.405±0.005 ampere."

What is the structure of inferences from indications to outcomes, and how can such inferences be justified from an epistemological perspective? Several challenges arise when one attempts to answer these questions. The remainder of this section (see "Objectivity and the Challenge of Context") will discuss two such challenges, the first one epistemological and the second methodological. The second section (see "Normativity and the Challenge of Practice") will argue that a model-based approach overcomes both challenges.[8]

Objectivity and the Challenge of Context

The first challenge to the epistemology of measurement arises from the fact that measurement outcomes are underdetermined by instrument indications. The same instrument indications may be taken as evidence for multiple, and in some cases inconsistent, knowledge claims about the values of the quantity being measured, depending on which background assumptions are used to interpret indications. A well-known example is the use of different fluids—such as alcohol, mercury, water, and air—to measure temperature (see Chang 2005: ch. 2).

Different fluids expand at different rates when heated, as do the different kinds of glass used as containers. If one assumes that each thermometer's volume indications are linearly correlated with temperature, their measurement outcomes turn out to be mutually inconsistent. That is, the same objects are assigned inconsistent temperature values when different thermometers are being used. However, if one assumes a more complex, nonlinear relation between the indications and outcomes of some of the thermometers, their outcomes can be made to agree without any change to indications. The choice of such nonlinear calibration functions often depends on background theoretical considerations, e.g. the assumption that air expands uniformly with temperature and can therefore serve as a standard. This simple and crude form of calibration already shows that measurement outcomes depend on one's background assumptions and on the abstractions and idealizations employed in the representation of the relevant measurement process, rather than on instrument indications alone. This point will be further reinforced by the more complex forms of calibration discussed below.

The sensitivity of measurement outcomes to representational context is an indispensable part of the practice of measurement. Measurement outcomes are often revised in light of new information about the instrument and the environment, or as a result of replacing some of the idealized assumptions involved in representing the measurement process with more realistic ones. As in the thermometer example, these revisions often do not involve any physical modification to the apparatus, the environment, or the measured object, but arise purely from altering the way indications are interpreted.[9] Without this context-sensitivity, methods of error correction and uncertainty evaluation that are habitually used in scientific laboratories would be impossible. Any epistemological account of measurement must accommodate sensitivity to representational context if it is to make sense of such common scientific practices.

At the same time, context-sensitivity seems to be at odds with another key desideratum for measurement outcomes, namely their objectivity. As mentioned, measurement outcomes are knowledge claims concerning the object or event being measured independently of the particular instruments and procedures used for its measurement, the particular environmental conditions in which it is measured, and the assumptions of those who measure it. Unless objectivity is established, there are no grounds for thinking that measurement outcomes obtained through the use of one kind of instrument in one sort of environment and under particular assumptions would still be valid under different circumstances and assumptions, and hence no grounds for comparing measurement outcomes to each other across laboratories. Moreover, without objectivity any talk of measurement accuracy makes very little sense. Any successful epistemological account of measurement must therefore clarify the conditions under which a measurement outcome may be justifiably deemed objective.[10]

How can measurement outcomes be both context-sensitive and objective? The difficulty in answering this question is what I call the "challenge of context." At first, meeting this challenge may appear impossible: it may seem like the two requirements are simply inconsistent. Below I will argue that the opposite is the case: the context-sensitivity of measurement outcomes is a necessary precondition for the possibility of establishing their objectivity. This becomes clear once a model-based epistemological account is adopted.

Normativity and the Challenge of Practice

A successful epistemology of measurement should do more than simply describe the methods scientists employ when they measure. It should also be able to provide insight into the sort of methods scientists *ought* to follow if measurement is to be a reliable source of knowledge. And yet one may rightly doubt whether philosophers are in a suitable position to make such normative judgments. Presumably, working scientists are more familiar than philosophers with the problems, methods, and equipment related to measurement in their field.

Philosophers therefore face two risks. On the one hand, they must take great care to ensure that their analyses of measurement do not lose touch with practice. If philosophers hope to make an informed contribution to debates concerning measurement methodology, they should first learn from practicing scientists which problems they are trying to solve, and which methods have proved most successful for dealing with these problems thus far. On the other hand, philosophers must not abandon their normative aims and confine themselves merely to reporting the methodological choices of scientists. Rather, philosophers are required to maintain a critical and reflective attitude towards practices of measurement. I call this dual requirement "the challenge of practice." This challenge applies to philosophy of science in general (and, indeed, to any philosophy of X where X is a domain of independently established human practice). But when it comes to measurement, the challenge of practice is associated with particular hazards.

As an example, consider the notions of measurement accuracy and measurement error. Philosophers with a metaphysical leaning have argued that the best way to explain scientists' attempts to increase measurement accuracy and diminish measurement error is by assuming the existence of true, mind-independent quantity values, which measurement outcomes are intended to approximate.[11] Setting aside debates about realism, the existence of true values of this sort cannot shed any light on the possibility of *evaluating* measurement accuracy and error. Even if such true values existed for some measurable quantity, they could only be approximated through measurement. Consequently, scientists could have no independent cognitive access to such true values and could not evaluate the accuracy and error of measurement outcomes against such values. Indeed, if scientists had access to such true values, measurement would be pointless, as the relevant quantity values would already be known with complete exactitude.

How, then, is the evaluation of measurement accuracy and error possible? At this point some philosophers may be inclined to embrace the idea that measurement accuracy and error, just like the true values of quantities, are unknowable in principle. Even measurement experts are tempted by such thoughts at times of metaphysical rumination (see JCGM 2012: Definition 2.13). While such views are admissible in a metaphysical debate, as part of an epistemology of measurement that aspires to make sense of scientific practice they are simply marks of terminological confusion. The terms "measurement accuracy" and "measurement error" have already established stable meanings set by their usage in scientific practice. Claims to evaluate measurement accuracy and error are habitually made in the sciences based on these stable meanings. Part of the task of epistemologists of measurement is to clarify these meanings and make sense of scientific claims made in light of such meanings. In some cases, the epistemologist may conclude that a particular scientific claim is unfounded or that a particular scientific method is unreliable. But the conclusion that *all* claims to accuracy and error are unfounded is only possible if philosophers ignore or misunderstand the established meanings of these terms.

Instead of comparing outcomes to true values, practicing scientists are faced with the challenge of evaluating accuracy and error by comparing measurement outcomes *to each other*. Such comparisons by their very nature cannot determine the extent of error associated with any single outcome but only overall mutual compatibility among outcomes. Multiple ways of distributing individual errors and uncertainties among outcomes are possible that are all consistent with the indications gathered from measuring instruments. Like measurement outcomes, then, claims to measurement accuracy and error are necessarily underdetermined by instrument indications. This predicament, which I have elsewhere called "the problem of accuracy" (Tal 2012: 11–2), is an epistemological difficulty with which scientists are actually faced, and for which they have devised a variety of solutions. Once philosophers acknowledge the existence and complexity of the problem of accuracy, they can devote efforts to study its sources, the conditions under which it can be solved, and the advantages and drawback of particular solutions. One of the findings of such study is that the term "measurement accuracy" carries multiple meanings in scientific practice, and that each meaning is associated with different normative constraints on the justification of accuracy claims (see Tal 2011).

As illustrated by this brief example, the challenge of practice is particularly hazardous for the study of measurement. Philosophical myths about measurement abound, and even seemingly innocuous presuppositions, like the idea that accuracy is a binary relation between an estimate and the truth, may turn out to be too simplistic to make sense of practice. To overcome the challenge of practice, epistemologists must not impose normative criteria that are so strict as to be unattainable. On the other hand, epistemologists also need to use their tools of abstraction and reflection to locate the sources of epistemological problems encountered in practice, and to critically assess the efficacy (or lack of efficacy) of particular methods used for their solution. Below I will show how a model-based conception of calibration solves the problem of accuracy, thereby shedding light on the prospects of tackling the challenge of practice.

Why Models?

The previous section clarified the goals of the epistemology of measurement and discussed two of its key challenges. This section will outline a model-based approach to the epistemology of measurement, and highlight the advantages of this approach for tackling both challenges. Parts of the discussion below summarize more detailed work published elsewhere (see Tal 2011; 2012; 2013).[12]

Coherence and Idealization

According to the account I will now propose, measurement is an activity aimed at the coherent and consistent attribution of values to one or more parameters in an idealized model of a process, based on the final states of that process.[13] Measurement involves at least two levels: the level of concrete interaction, or process, and the level of abstract representation, or model. The model is usually constructed from theoretical assumptions concerning the relevant process as well as statistical assumptions concerning the data generated by that process. The physical process itself includes all actual interactions among measured objects, instrument, operators and environment, but the models used to represent such processes neglect or simplify many of these interactions. It is only in light of such idealized model, I will argue, that measurement outcomes can be assessed for accuracy and meaningfully compared to each other. Indeed, it is only against the background of such simplified and approximate representation of the measuring process that measurement outcomes can be considered candidates for objective knowledge.

These claims may seem counterintuitive. How can idealization and approximation, which are the results of distorting or neglecting aspects of reality, be necessary preconditions for the justification of claims to accuracy and objectivity? To answer this question, it is useful to return to the distinction between instrument indications and measurement outcomes. Instrument indications do not by themselves provide any objective knowledge, that is, knowledge about the objects intended to be measured. This is because indications are the products of a complex interaction between the object intended to be measured, the instrument, the environment and the humans performing the measurement (the "operators"). Variation among indications may reflect pertinent differences among the objects intended to be measured, but they may also reflect variations in the way the instrument operates and interacts with the objects, environment, and operators. Grounding knowledge claims about the state of an object in isolation from the instrument and environment requires differentiating the two kinds of variation.

Such differentiation, however, cannot be accomplished simply by observing the statistical or algebraic properties of indications. As an example, consider again the case of comparability among different thermometric fluids. Temperature intervals deemed equal by a thermometer filled with one kind of fluid are deemed unequal by a thermometer filled with another (see Chang 2005: 57–60). These discrepancies remain stable over repeated runs, and are therefore statistically significant. Moreover, the expansion rates of different fluids are nonlinearly related and hence cannot be eliminated through linear scale transformations. Given nothing but the statistical and algebraic properties of indications, it appears that one would have to conclude that the instruments are measuring different kinds of quantity.

The only way of avoiding such extreme operationalism without physically modifying the instruments is by representing each instrument in a manner that idealizes away its idiosyncrasies. In our case, this involves constructing calibration functions that map the indications of each instrument to consistent temperature values. These calibration functions idealize away differences in the rates of thermal expansion of each fluid as well as the effects of thermal expansion on the containers. The instruments are properly deemed to be in agreement when such idealized representations provide converging value assignments to a common parameter—namely, temperature—under relevantly similar circumstances. More generally, different measurement processes provide objective knowledge about the values of a quantity only once they have been *idealized in a mutually coherent and consistent manner* in terms of that quantity. Measurement outcomes are attributions of values or value-ranges to one or more parameters in such idealized models, rather than claims about observable relations among measured objects themselves.[14]

A direct upshot of the foregoing discussion is that measurement outcomes need not, and indeed usually do not, preserve the algebraic structure of indications. The outcomes associated with a properly calibrated alcohol thermometer, for example, deem two temperature intervals equal just in case its indications deem them *unequal*.[15] The same holds true for clocks. Any two clocks, when compared, will eventually drift in frequency relative to each other. Consequently, for any pair of clocks, there exists a series of time intervals such that one clock orders their durations in a reverse way to the other. Nonetheless, when two such clocks are modeled in a manner that idealizes away their relative drift, they provide a consistent ordering of those time intervals. As I have shown elsewhere, such idealized models are essential to the modern standardization of time (see Tal 2016).

Contrary to the widely held Representational Theory of Measurement (RTM) (Krantz et al. 1971), then, measurement is not the establishment of homomorphisms between qualitative and quantitative relational structures. Whenever the indications of instruments that are intended to measure the same quantity display systematic and nonlinear discrepancies, measurement outcomes must be adjusted in a manner that no longer preserves order and equality among indications, let alone more complex structures like additivity. Given that such discrepancies are the rule rather than the exception in scientific practice, the view that structure-preservation is necessary for measurement is mistaken. The mistake arises from the naive empiricist confusion between the mathematical and epistemological foundations of measurement, and particularly from the assumption that the mathematical structures used in measurement must roughly reflect directly observable structures. Once this assumption is dropped, RTM is more charitably understood as a theory of quantification or scale-construction than a theory of measurement.[16]

Calibration as a Modeling Activity

A crucial step in establishing the objectivity of measurement outcomes is the construction of a calibration function. A calibration function maps instrument indications to measurement outcomes, thereby allowing scientists to distinguish the pertinent information conveyed by indications from local and idiosyncratic features of instruments, environments, and operators. How are calibration functions constructed, and what criteria constrain their selection?

As I will argue in this section, calibration is a special sort of modeling activity, one in which the system being modeled is a measurement process. I propose to view calibration as a modeling activity in the full-blown sense of the term "modeling," i.e. constructing an abstract and idealized representation of a system from theoretical and statistical assumptions and using this representation to predict that system's behavior. Calibration activities are often iterative and involve modifications to both the measurement process and the model representing that process. Calibration functions are accordingly *predictions* derived from such models, and the choice of calibration function for a given instrument is constrained by the requirements of coherence and consistency among such models.

The simplest case of calibration is known as "black-box" calibration.[17] Here the measurement process is modeled as an input-output device: quantity value in, indication out. None of the secondary influences on the relationship between quantity values and indications are explicitly represented. This mode of calibration typically proceeds by collecting a set of indications for quantity values that are already known and fitting a curve through them. That curve is then predicted to hold when the instrument interacts with objects whose relevant quantity values are not yet known, allowing scientists to draw inferences in the inverse direction. For example, a simple caliper may be calibrated by placing gauge-blocks of known lengths between its jaws, fitting a linear function to the resulting indications, and using this function to predict the response of the caliper to unknown lengths.

As straightforward and theory-free as it may seem, black-box calibration already exhibits most of the characteristics of calibration mentioned above. First, it involves the construction of a simplified model of the instrument and its interactions with the objects intended to be measured. Second, this model involves theoretical assumptions, e.g. that the variation among indications is functionally related to the variation of lengths, and that this functional relation is stable. Third, the model involves a statistical idealization of the instrument's behavior, namely, the use of a simple and smooth mathematical curve to approximate its response to standard lengths. And fourth, this simplified model of the instrument is used to predict which quantity values are likely to be associated with particular indications in future applications.

Black-box calibration yields a simple but inaccurate model of the relevant measurement process. The model neglects extrinsic influences on the production of indications, thereby limiting one's ability to predict how indications will change when those influencing factors change. As a result, measurement outcomes generated through black-box calibration are especially prone to systematic errors. By contrast, white-box calibration yields a complex but more accurate model. White-box calibration procedures represent the measurement process as a collection of modules, rather than as a single input/output unit. Each module is characterized by one or more state parameters, laws of temporal evolution, and laws of interaction with other modules. The collection of modules and laws constitutes a more detailed (but still idealized) model of the measurement process than a black-box model.

For example, the process of operating a caliper to measure diameter can be represented by a collection of modules (see Schwenke et al. 2000: 396), such as the caliper's jaws, its scale, the workpiece whose diameter one intends to measure, the area of contact between the caliper and the workpiece, and the caliper's readout.[18] Each module is associated with one or more parameters, e.g. the roughness of the contact, the resolution of the readout, the temperature of the workpiece, and so on. These parameters are then assumed to enter into certain functional relations with each other, as well as with other parameters such as time, in light of background theories and statistical assumptions. These functional relations are typically expressed as a set of equations, and solving these equations yields a calibration function, that is, a prediction as to which diameters will be correlated with which caliper indications. This prediction is tested against gauge-blocks or other objects whose lengths are already known, before being projected into the future and used to infer unknown diameters from the indications of the caliper.

Thanks to its detailed representation of the structure and dynamics of the measurement process, white-box calibration usually results in higher measurement accuracy than black-box calibration. As systematic biases are explicitly represented and corrected, the final uncertainty of outcomes tends to be smaller than in the black-box case. White-box calibration also provides clearer limits on the tolerable variations to each operational and environmental factor that affects the instrument's indications. This is because the white-box evaluation of measurement uncertainty is obtained by propagating uncertainties from input to output parameters in the model (JCGM 2008), a method that involves explicit assumptions about the probability distributions of each operational and environmental factor.

The Myth of Absolute Accuracy

The picture of calibration I have outlined thus far differs from the traditional way of viewing calibration in philosophy and metrology. Although the centrality of modeling to some kinds of calibration is widely acknowledged, calibration is traditionally defined as the activity of establishing a correlation between the indications of a measuring instrument and quantity values associated with a measurement standard (see Boumans 2007: 236).[19] This definition, I will now argue, provides only a partial and limited understanding of the epistemic efficacy of calibration procedures. In particular, it mischaracterizes the reasons that calibration is successful in evaluating measurement accuracy.

At first, the traditional definition of calibration may seem easy to vindicate: if we suppose that measurement standards are absolutely accurate by definition, it immediately follows that the activity of finding correlations between the indications of an instrument and quantity values associated with standards provides knowledge about the instrument's accuracy. However, the premise clashes with scientific practice. In practice, measurement standards are not considered to be absolutely accurate, but are assumed to suffer from uncertainties like any other material procedure. Moreover, although the uncertainties associated with measurement standards are usually very small compared to most other measurements of the same quantity, this is not primarily due to any definition. On the contrary, metrologists choose to use certain procedures as standards because those procedures have already been shown to be more accurate than previous standards (see Tal 2011: 1093–4).

In order to make sense of the practice of measurement, the myth of the absolute accuracy of measurement standards needs to be re-examined and its sources clarified. The myth results from confusing the definition of a measurable quantity, which is a linguistic entity, with the *realizations* of that definition, which are concrete procedures for approximately satisfying the definition in a replicable manner. The clause "in a replicable manner" is crucial, as metrologists are not simply interested in fixing the meanings of terms but in creating a universally stable system of measurement. In what follows the term "measurement standard" will refer to realization rather than to definition.

Measuring instruments cannot be calibrated against definitions, but only against realizations, which are always somewhat inaccurate. Even when a definition makes explicit reference to an aspect of a particular realization, that realization is not thereby assigned an uncertainty of zero. An example is the current definition of the kilogram as the mass of the International Prototype of the Kilogram (IPK). The metal object designated "IPK" is only one part in the realization of the kilogram. In addition to IPK, the replicable satisfaction of the definition of the kilogram includes a variety of maintenance and cleaning procedures, sensitive balances for comparing the IPK with other masses, and theoretical and statistical models for analyzing and correcting the results of such comparisons (see Girard 1994). Like any set of physical procedures, the replication of the kilogram is not completely exact and therefore involves some uncertainty. This uncertainty represents the degree of unpredictability associated with determinations of mass relative to that of IPK, and is evaluated experimentally. Metrologists are continually improving other techniques for measuring mass, such as watt balances, with the aim of surpassing the accuracy of mass comparisons against the IPK and eventually replacing the current definition of the kilogram with a definition that is more accurately realizable.[20]

The Epistemic Role of Standards

The fact that measurement standards are not absolutely accurate poses a problem for the traditional conception of calibration. Why should scientists calibrate measuring instruments against standards rather than against any other, sufficiently accurate measuring procedure? In other words, is there anything epistemically special about the quantity values supplied by the procedures metrologists call "standards?" My answer will be a qualified "no." As long as one is concerned with local, pairwise comparisons between instruments, it makes no epistemic difference whether (or which) one of the instruments is designated a "standard." The total uncertainty associated with the values being compared remains the same, and is arrived at through the same chain of inferences, regardless of such designation. The epistemic difference associated with the title "standard" appears only on a global scale, when metrologists are required to distribute uncertainties across large networks of instruments.

Let us begin with the local context, and consider the calibration of a measuring instrument against a metrologically-sanctioned standard. For example, consider a scenario in which a cesium fountain clock—an official realization of the definition of the second—is used to calibrate a hydrogen spectrometer, i.e. a device for measuring the frequency associated with subatomic transitions in hydrogen.[21] The accuracy expected of the spectrometer is close to that of the standard, so that one cannot neglect the inaccuracies associated with the standard during calibration. The inaccuracies of measurement standards, however, are evaluated through the same principles and procedures that were already noted for the calibration of non-standard measuring procedures. The accuracy of a cesium fountain clock, for example, is determined by constructing a white-box model of the instrument and its environment. This model is used to derive a function that maps the raw output frequency of the clock (its indications, or "ticks") with the frequency associated with a transition between two states of a cesium atom under ideal conditions (the measurement outcome). This idealized frequency is currently used to define the second.[22] The accuracy of the cesium clock is evaluated by propagating uncertainties through this white-box model, and then comparing the consequences of the model to those derived from models of other atomic clocks.[23] In other words, the designation "standard" does not alter the way accuracy is evaluated in pairwise comparisons between clocks.

When a hydrogen spectrometer is calibrated against a standard cesium clock, metrologists do not compare the raw frequencies generated by each instrument. Such comparison between indications would be uninformative, as each instrument has different idiosyncrasies and is therefore affected by different biases. Instead, the informative comparison is between the consequences of two white-box models, one for the spectrometer and one for the standard clock. Both the spectrometer and the standard clock are deemed accurate when the measurement outcomes assigned by their respective models maintain a stable relation with each other, as predicted by the relevant background theories. In our example, this means that the parameter values assigned to the ideal atomic transition frequencies of hydrogen and cesium based on the indications of each instrument have a stable ratio. Unlike the ratio between indications, which is a local property of two instruments and is influenced by various operational and environmental factors, the ratio between model parameter values is isolated from intervening factors and can therefore form the basis for an objective knowledge claim about the relevant atomic properties.

This two-way white-box comparison exemplifies calibration in its full generality and exposes the limitations of the traditional conception of calibration. In its full generality, *calibration is the activity of modeling different processes and testing the consequences of such models for mutual compatibility.* Such activity is indifferent to whether or which of the processes are designated "standards," and indeed in some domains of physical measurement one encounters calibration without standards.[24]

Contrary to the traditional characterization of calibration, then, comparison with a standard is neither necessary nor sufficient for successful calibration. The designation "measurement standard" does not guarantee that tests for compatibility among model consequences will succeed; on the contrary, this designation is given after the fact to certain processes that metrologists have already succeeded in modeling in a coherent manner. Comparison to a metrologically-sanctioned standard is therefore only a proxy method for testing for compatibility among models. Indeed, only in the simplest and most inaccurate case of calibration (black-box calibration) is compatibility achieved simply by establishing empirical correlations between instrument indications and standard values.

The traditional conception of calibration mistakes this simplest form of calibration for the general case. The opposite is true: black-box calibration is but a special case of white-box calibration in which there are only two modules (instrument and object) and two parameters (indication and outcomes). One-way white-box calibration is in turn a special case of two-way white-box calibration in which the uncertainties associated with one of the instruments are neglected.

One may attempt to save the traditional conception of calibration by using the term "standard" more broadly to include any accurate measuring procedure, rather than only metrologically-sanctioned realizations. However, the traditional conception of calibration is not merely too narrow; it also misidentifies the source of epistemic efficacy of calibration operations. The primary epistemic function of calibration—evaluating accuracy—is achieved by evaluating coherence among models, not by evaluating empirical correlations among the indications of instruments.[25] From a methodological point of view, the accuracy of a measurement is nothing but the predictive accuracy of the model used to represent the relevant measurement process, and calibration is successful in evaluating measurement accuracy because it cross-checks such model predictions against each other. The traditional focus on standards obscures this underlying inferential structure and misses the point behind instrument comparisons.

Nonetheless, the traditional conception of calibration is not completely wrong. The designation "standard," although not necessary for calibration, is often extremely useful. This is because the accuracy of individual measuring instruments is underdetermined by empirical evidence from comparisons, a predicament that has already been discussed above under the title "the problem of accuracy." The mutual predictability of two instruments' indications under their respective models only determines their *joint* accuracy. To isolate individual accuracies, a web of comparisons is required that would determine which instrument is better at predicting the behaviors of yet *other* instruments, again under their respective models. However, the web of comparisons cannot completely escape the underdetermination problem. This becomes especially clear when the network of instruments is large enough to break into "islands," that is, subsets of measuring procedures that maintain good mutual predictability among themselves but support only inaccurate predictions of the indications of procedures in other "islands."

As an example, consider the ensemble of commercial atomic clocks that is currently used to standardize global timekeeping (known as the EAL ensemble). Clocks in this ensemble usually "tick" at a very stable rate relative to each other, but occasionally some of them drift in frequency relative to the ensemble average. In order to distribute uncertainties consistently across the web of instruments, the scientific community must make a partially conventional choice as to which of those instruments are to be considered more accurate. A useful method of regulating the distribution of uncertainties is by labeling a particular subset of measurement procedures "standards," and assigning as much uncertainty as practically possible to *other* measurement procedures whenever disagreements occur. Whenever a clock's frequency veers too much from the ensemble average, metrologists deem that clock unstable and ignore its indications in an *ad hoc* fashion, essentially demoting it from the status of timekeeping standard, although they could have just as reasonably attributed the instability to the ensemble average itself.[26]

This brief example suggests that the designation "standard" does after all affect the accuracy of a procedure thus designated, albeit only in a limited way. In their ongoing attempts to maintain both accuracy and reproducibility among measurement outcomes, metrologists bestow the title "standard" to those processes and procedures whose use as references would minimize uncertainty without breaking coherence. The regulation of uncertainty assignments throughout a large network of instruments so as to maintain overall coherence is one of the central epistemic functions of standardization, as I have discussed elsewhere.[27] The policies that govern such regulation practices have an element of convention, but at the same time are also empirically constrained by the requirement to maintain reproducibility across a large web of comparisons. Calibration against standards is accordingly a highly useful practice from a global perspective, when one is concerned with the integrity of entire metrological networks.

Robust Perspectivism

What is being compared when measuring instruments are compared? As I have argued, it is not the instruments' indications, but rather the values of parameters in an idealized model that predicts those indications, which are the proper loci of comparison. Evaluating measurement accuracy is tantamount to evaluating the predictive accuracy of such a model. The direct consequence of these insights is that measurement outcomes are context-sensitive and relative to an abstract and idealized representation of the procedure by which they were obtained. This explains how the outcomes of a measurement procedure can change without any physical modification to that procedure, merely by changing the way the instrument is represented (recall the seemingly divergent thermometers). Second, it explains how the accuracy of a measuring instrument can be improved merely by adding detail to the model representing the instrument, for example, by replacing a black-box model with a white-box model. Third, the model-relativity of measurement outcomes explains how the same set of operations, again without physical change, can be used to measure different quantities on different occasions depending on the interests of researchers.[28]

The context-sensitivity of measurement outcomes in the model-based account does not hinder their objectivity. Indeed, the context-sensitivity of measurement outcomes is a necessary precondition for their obtaining objective status. For a putative measurement outcome to be deemed objective, it must satisfy the requirement of *convergence under multiple models*. This requirement decreases the likelihood that measurement outcomes would be mere artefacts of local, theoretical or statistical assumptions. More importantly, the requirement of convergence under models is necessary for representing multiple instruments as measuring the *same* quantity. Prior to their representation by an idealized model, there is no way of testing whether different instruments measure the same quantity; any agreement or disagreement among their indications may be construed as coincidental and attributed to some local feature of the instruments or environments. It is only once their idiosyncrasies are idealized away in a mutually coherent fashion that instruments can be viewed as sources of objective knowledge about a common quantity, such as temperature or frequency.

Some philosophers may desire a stronger sense of objectivity than the one I have employed above. They may wish to ground certain claims about measurement, such as the claim that the expansion of mercury measures temperature, on perspective-independent truths. As I have already clarified at the beginning of this contribution, such demands set the epistemic bar too high, and in so doing risk misunderstanding the actual problems scientists face when designing and calibrating measuring instruments.

In place of perspective-independence, I propose to think of objectivity in measurement as perspective-invariance, namely as the robustness of measurement outcomes across different material circumstances and representational contexts. According to this *robust perspectivism*, measurement outcomes need not (and cannot) be independent of any context whatsoever. At the same time, measurement outcomes should not depend on any *particular* context. Robust perspectivism has the double advantage of setting strict normative constraints on the kind of knowledge claims that may be deemed objective, while being compatible with the methodological precepts scientists actually follow in exemplary cases of knowledge production.

By embracing robust perspectivism, a model-based epistemology of measurement is in a good position to tackle both the challenge of context and the challenge of practice. The first challenge is met by reconciling the context-dependence of measurement outcomes with their objectivity, as already noted. The second challenge is met, on the one hand, by requiring epistemologists to engage with the problems and methods that practicing scientists identify as central to measurement in their discipline; for example, the problem of evaluating accuracy by comparing multiple inaccurate measurement outcomes to each other. On the other hand, epistemologists are required to formulate general normative criteria that extend beyond concrete methods and problems. In some cases, this may lead epistemologists to question claims made by practitioners.

In the case of metrological calibration, for example, the general solution to the problem of accuracy turned out to be the normative criterion of convergence under models. This solution has led to a critique of the traditional definition of calibration as formulated by practitioners themselves. Pursued in this manner, the epistemology of measurement promises to be neither an armchair activity devoid of contact with experience, nor a purely descriptive science, but a reflective discipline charged with the explication and assessment of the concepts and methods used in measurement.

Acknowledgments

The author would like to thank the participants of the *Dimensions of Measurement* conference, held in Bielefeld in March 2013, for valuable feedback. I am grateful to Nicola Mößner and Luca Mari for helpful comments on a draft version of this contribution. This research was supported by an Alexander von Humboldt Postdoctoral Research Fellowship and a Marie Curie Intra-European Fellowship within the 7th European Community Framework Programme.

Notes

1 The category "concrete physical process" should be understood broadly, and may include, for example, computer hardware running a simulation algorithm or a human subject interacting with a questionnaire.
2 Most notably in Chang (2005: ch. 5) and van Fraassen (2008: ch. 5–7).
3 JCGM (Joint Committee for Guides in Metrology), *International Vocabulary of Metrology—Basic and General Concepts and Associated Terms* (VIM), 3rd edition

with minor corrections (Sèvres: JCGM, 2012). http://www.bipm.org/en/publica
tions/guides/vim.html, definition 2.2.

4　The difference between numbers and numerals is important here. Before processing, an indication is never a number, although it may be a numeral (i.e. a symbol
representing a number).

5　Quantitative measurement outcomes are often recorded in the form of a best
estimate and a standard deviation that represents the width of a normal probability distribution around the mean, but other forms are also commonly used,
e.g. min-max value range without a best estimate, or an asymmetrical probability
distribution. For a clarification of my use of the term "estimate" see fn 7.

6　The position of the ammeter pointer may itself be measured, of course, but in
that case one would be measuring the ammeter rather than *with* the ammeter.
More generally, it is common for inferences from indications to (final) measurement outcomes to involve intermediary measurement outcomes. Wendy Parker
distinguishes among three kinds of measurement in accordance with their levels
of inferential complexity, which she calls "direct," "derived," and "complex"
(Parker 2015).

7　Talk of "best estimates" is not intended to imply that measurable quantities
have true, mind-independent values, but only that measurement involves uncertainty and is therefore usually subject to improvement. As will be clarified below,
measurement uncertainty is a special case of predictive uncertainty. Hence reducing measurement uncertainty amounts to reducing the uncertainty with which a
model of the measurement process predicts that process' behavior. Measurement
outcomes may be viewed as estimates insofar as they presuppose the possibility of
further improvement to the predictive accuracy of the relevant models.

8　These are not the only challenges facing the epistemology of measurement.
Another kind of challenge that has received significant attention in recent years
is the problem of coordination (or the "problem of nomic measurement") (see
Chang 2005: ch. 2; van Fraassen 2008: ch. 5). Other challenges include the problem of quantity individuation (see Tal 2012: ch. 2) and the problem of observational grounding (see Tal 2013: 1167–8). Space limitations prevent discussion of
these problems here.

9　For an example of error correction through de-idealization see Tal (2011: 1082–96).

10　"Objective" here does not imply "real." The precise sense of objectivity employed
in this contribution will be clarified below in the section on robust perspectivism.

11　See, for example, Swoyer (1998: 239) and Trout (1998).

12　For closely related model-based approaches to measurement see Mari (2005:
259–66); Frigerio et al. (2010: 123–49); Giordani and Mari (2012: 2144–52); Mari
and Giordani (2013: ch. 4).

13　This is not meant as a definition of measurement, but only as a characterization of
what the activity of measuring is intended to accomplish.

14　The distinction between indications and outcomes echoes the distinction between
data and phenomena drawn by Bogen and Woodward, although the two distinctions do not exactly overlap (Bogen and Woodward 1988).

15　This assumes that the thermometer's display has not yet been adjusted to conceal
the nonlinear correction.

16　Interpreted as a theory of scale-construction, RTM remains useful for identifying the assumptions (known as "axioms") involved in choices of measurement
scale, but is no longer expected to clarify how those axioms are justified in light of
empirical evidence. For discussion see Tal (2012: ch. 2).

17　For the distinction between black-box and white-box calibration see Boumans
(2006; 2007).

18　One can easily imagine more or less detailed breakdowns of a caliper into modules than the one offered here. The term "white-box" should be understood as referring to a wide variety of modular representations of the measurement process with differing degrees of complexity, rather than a unique mode of representation. Simple modular representations are sometimes referred to as "grey-box" models (see Boumans 2006: 121–2).

19 See also the *International Vocabulary of Metrology* (JCGM 2012: 2.39). Franklin characterizes calibration more broadly as a reproducibility test that employs a "surrogate signal," which need not be an official measurement standard (see Franklin 1997). For a discussion of these views see Soler et al. (2013).

20 Metrologists plan to redefine the kilogram in 2018 by fixing the value of the Planck constant (see Milton, Davis, and Fletcher 2014).

21 Such calibration is described in Niering et al. (2000: 5496).

22 Since 1967 the second has been defined as "the duration of 9,192,631,770 periods of the radiation corresponding to the transition between the two hyperfine levels of the ground state of the caesium 133 atom" (BIPM (International Bureau of Weights and Measures) 2014 [2006]: 113).

23 For a more detailed account of the accuracy evaluation of primary frequency standards see Tal (2011).

24 An example is the measurement of industrial benchmarks such as paper quality (see Wirandi and Lauber 2006).

25 A precise characterization of the kind of coherence required is given by the Robustness Condition discussed in Tal (2011: 1091; 2012: 175).

26 For a detailed discussion of this example see Tal (2016).

27 See Tal (2016) for discussion.

28 An example is the use of the same pendulum to measure either duration or gravitational potential without any physical change to the pendulum or to the procedures of its operation and observation. The change is effected merely by a modification to the mathematical manipulation of quantities in the model. For measuring duration, researchers plug in known values for gravitational potential in their model of the pendulum and use the indications of the pendulum (i.e. number of swings) to tell the time, whereas measuring gravitational potential involves the reverse mathematical procedure.

References

BIPM (International Bureau of Weights and Measures). *SI Brochure: The International System of Units*, 8th edn. 2014 [2006], http://www.bpm.org/en/publications/si-brochure/, accessed March 17, 2016.

Bogen, J. and Woodward, J. 1988. Saving the Phenomena. *Philosophical Review* 97:3: 303–52.

Boumans, M. 2007. Invariance and Calibration. In: *Measurement in Economics: A Handbook*, edited by M. Boumans, 231–48, Amsterdam: Elsevier.

Boumans, M. 2006. The Difference Between Answering a "Why" Question and Answering a "How Much" Question. In: *Simulation: Pragmatic Construction of Reality*, edited by J. Lenhard, G. Küppers, and T. Shinn, 107–24, Dordrecht: Springer.

Boumans, M. 2005. *Economists Model the World into Numbers*. New York: Routledge.

Bridgman, P. W. 1927. *The Logic of Modern Physics*. New York: Macmillan.

Campbell, N. R. 1920. *Physics: The Elements*. London: Cambridge University Press.

Carnap, R. 1995 [1966]. Philosophical Foundations of Physics. In: Carnap, R., *An Introduction to the Philosophy of Science*, edited by M. Gardner, New York: Dover.

Chang, H. 2004. *Inventing Temperature: Measurement and Scientific Progress*. Oxford: Oxford University Press.

Duhem, P. 1962 [1906]. *The Aim and Structure of Physical Theory*, translated by P. P. Wiener. New York: Atheneum.

Ellis, B. 1966. *Basic Concepts of Measurement*. Cambridge: Cambridge University Press.

Franklin, A. 1997. Calibration. *Perspectives on Science* 5:*1*: 31–80.

Frigerio, A., Giordani, A., and Mari, L. 2010. Outline of a General Model of Measurement. *Synthese* 175: 123–49.

Giordani, A. and Mari, L. 2012. Measurement, Models, and Uncertainty. *IEEE Transactions on Instrumentation and Measurement* 61:*8*: 2144–52.

Girard, G. 1994. The Third Periodic Verification of National Prototypes of the Kilogram (1988–1992). *Metrologia* 31: 317–36.

Gooday, G. J. N. 2004. *The Morals of Measurement: Accuracy, Irony, and Trust in Late Victorian Electrical Practice*. Cambridge: Cambridge University Press.

JCGM (Joint Committee for Guides in Metrology), *International Vocabulary of Metrology—Basic and General Concepts and Associated Terms* (VIM), 3rd edition with minor corrections. 2012, http://www.im.org/en/publications/guides/vim.html, accessed March 17, 2016.

JCGM (Joint Committee for Guides in Metrology), *Guide to the Expression of Uncertainty in Measurement* (Sèvres: JCGM, 2008), http://www.bipm.org/en/publications/guides/gum.html, accessed March 17, 2016.

Krantz, D. H., Suppes, P., Luce, R. D., and Tversky, A. 1971. *Foundations of Measurement Volume 1: Additive and Polynomial Representations*. New York: Dover.

Mari, L. and Giordani, A. 2013. Modeling Measurement: Error and Uncertainty. In: *Error and Uncertainty in Scientific Practice*, edited by M. Boumans, G. Hon, and A. Petersen, ch. 4, London: Pickering and Chatto.

Mari, L. 2005. The Problem of Foundations of Measurement. *Measurement* 38: 259–66.

Michell, J. 1994. Numbers as Quantitative Relations and the Traditional Theory of Measurement. *British Journal for the Philosophy of Science* 45: 389–406.

Milton, M. J. T., Davis, R., and Fletcher, N. 2014. Towards a New SI: a Review of Progress Made Since 2011. *Metrologia* 51: R21–R30.

Niering, M., Holzwarth, R., Reichert, J., Pokasov, P., Udem, Th., Weitz, M., Hänsch, T. W. et al. 2000. Measurement of the Hydrogen 1 S-2 S Transition Frequency by Phase Coherent Comparison with a Microwave Cesium Fountain Clock. *Physical Review Letters* 84:*24*: 5496.

Parker, W. (2015). Computer Simulation, Measurement and Data Assimilation. *British Journal for the Philosophy of Science*. doi: 10.1093/bjps/axv037.

Reichenbach, H. 1958 [1927]. *The Philosophy of Space and Time*. New York: Dover.

Schwenke, H., Siebert, B. R. L., Waldele, F., and Kunzmann, H. 2000. Assessment of Uncertainties in Dimensional Metrology by Monte Carlo Simulation: Proposal for a Modular and Visual Software. *CIRP Annals—Manufacturing Technology* 49:*1*: 395–8.

Soler, L.; Wieber, F.; Allamel-Raffin, C.; Gangloff, J. L.; Dufour, C., and Trizio, E. 2013. Calibration: A Conceptual Framework Applied to Scientific Practices Which Investigate Natural Phenomena by Means of Standardized Instruments. *Journal for General Philosophy of Science* 44:*2*: 263–317.

Swoyer, C. 1987. The Metaphysics of Measurement. In: *Measurement, Realism and Objectivity*, edited by J. Forge, 235–90, Dordrecht: Reidel.

Tal, E. 2016. Making Time: A Study in the Epistemology of Measurement. *British Journal for the Philosophy of Science* 67:*1*: 297–335.

Tal, E. 2013. Old and New Problems in Philosophy of Measurement. *Philosophy Compass* 8:*12*: 1159–73.

Tal, E. 2012. *The Epistemology of Measurement: A Model-Based Account*. PhD thesis. University of Toronto.

Tal, E. 2011. How Accurate Is the Standard Second? *Philosophy of Science* 78:*5*: 1082–96.

Thomson, W. 1891. Electrical Units of Measurement. In: *Popular Lectures and Addresses*, vol. 1, 25–88, London: McMillan.

Trout, J. D. 1998. *Measuring the Intentional World: Realism, Naturalism, and Quantitative Methods in the Behavioral Sciences*. Oxford: Oxford University Press.

Van Fraassen, B. C. 2008. *Scientific Representation: Paradoxes of Perspective*. Oxford: Oxford University Press.

Wirandi, J. and Lauber, A. 2006. Uncertainty and Traceable Calibration—How Modern Measurement Concepts Improve Product Quality in Process Industry. *Measurement* 39: 612–20.

Index

Figures indexed in bold

Academy for the Advancement of
  Science 178
accuracy 136, 138–40, 190–1, 194–5,
  223, 225, 229, 234, 238–40, 244–9;
  absolute 243–4; of angular distances
  191; of blood pressure outcomes
  136; epistemic 139–40; and error
  of measurement outcomes 238;
  evaluating 239, 246, 249; high
  mathematical 199; of individual
  measuring instruments 247; of mass
  comparisons 244; of medieval
  astrolabes 191; metaphysical 140;
  myth of absolute 243; of physical
  quantities 140; predictive 246, 248;
  representational 3; and reproducibility
  247; unprecedented 195
Actinograph slide rule **45**
actinographic discovery of the "chemical
  rays" of ultraviolet 44
active realism 2, 32, 34–6
aesthetic engagement 57
aesthetics 57–8; *see also* media
  aesthetics
Agassiz, Louis 13
Almagest 190–1, 193
alteration of MPIs and combinations 97
Amann, Klaus 88, 90, 92
American plant ecology 174
analysis 81–3
Analysis; canonical correlation 161;
  Fourier 222; Friedman 161; Kepler
  196; statistical 159, 207
Antares (star) 99, 102, 104
Arago, Dominique François 41–4, 46,
  48, 53
Archer, Scott 48

argument 19, 58–60, 62, 64, 67, 119–20,
  147, 151, 204, 206–8; convincing
  70; general 59, 68; parallel 68;
  visual 112
Aristarchus of Samos 190
Arizona 170–1, 179, 181
artefacts 44, 46, 48, 104, 112, 115, 117,
  119, 124, 203
astronomy 26, 81–2, 190–2, 194–7, 199
atomic clocks 138, 245, 247
Augustine of Hippo 113
automatic, or self-registration of nature
  42
autoradiographs 87–91

background knowledge 4, 101, 104, 107;
  relevant 235; scientist's 103, 107
Backus, George 225–6
Backus-Gilbert resolution method 226
Bacon, Roger 113
Bak, Per 84
Barad, Karen 21
Baraniuk, Richard 78
basic research 84
al-Battani 191
Beam, Lura E. 157
Beer, Gillian 19–20
Benjamin, Walter 46
Bennett, James 193
Bernard, Claude 46, 190
bias 135–8, 144, 245; gravitational
  139; measurement 137–9;
  systematic 243
big data science 3, 81
Bisexual Option 157, 159, 166
bisexuality 157–61, 165; psychological
  157; self-identified 159

Black, Max 115
blind spots 147, 149
blood pressure 43, 133–6, 138,
    141–2; accuracy of readings 136;
    conceptualization of measurement
    accuracy in 138; high 136; measured
    141; outcomes 136; systolic 142; true
    136, 138
Blow-Up 82, 84
Bogen, James 204, 206–9, 213
Boon, Mieke 5–6, 203–16
Borsboom, Denny 149–50
Boumans, Marcel 220–1, 228, 243
Bowers, Janice E. 171, 174, 177, 179
Brahe, Tycho 194, 198
brain imaging 57–9, 61–70, 79; cognitive
    70; common 68; contemporary 67;
    data 59, 63–5; public perceptions of
    67; techniques 63, 68; technologies
    57, 65
brain scans 3, 64–5
Brainbow mouse 63
Bridgman, Percy Williams 2, 25–35,
    133, 233; operationalism 28–9;
    philosophical writings 27; and the
    terms instrumental, paper-and-pencil
    and mental operations 31
Brock, Emily K. 5, 170–82
Bullen, Keith 220, 222
Burbank, Luther 173, 176

calibration 6, 14, 103, 107, 152, 234,
    236–7, 239, 241–7, 249; activities 242;
    black-box 242–3, 246; functions 236,
    241–3; metrological 234, 249; mistakes
    246; of nonstandard measuring 245;
    procedures 234, 242–3; white-box
    242–3, 246
California 157, 162, 173, 176
California Institute of Technology
    (Caltech) 25, 222
Campbell, Norman R. 1, 151, 233
Cannon, Susan Faye 16–17
canonical correlation analysis 161
Caplan, Jane 69
Carmel 176–9, 181
Carmel Development Company 176
Carnegie Coastal Laboratory 176, 180–1
Carnegie Desert Laboratory 5, 170–1
Carnegie-funded ecologists 178
Carnegie Institution of Washington 170,
    172–4, 176, 179, 181
cartography 16, 118

Cartwright, Nancy 30
Castel, Alan D. 57, 59–61, 63–4
CAT 74, 81
causality 83, 117, 234; often suspended
    in favour of a more heuristic and
    empirical view of the sciences
    84; and rules for translation or
    transformation 2; theory of
    correspondences by relations of 117
Cavell, Stanley 12
Chang, Hasok 25–36, 143, 152, 212, 233,
    236, 240
charity, principle of 123
Chura, Patrick 19–20
CMS 81
cognitive change 189, 198–9
cognitive integration 5, 189, 191, 193,
    195, 197–9
coherence 36, 90–1, 114, 119, 240, 242,
    247; evaluating 246; norms of 90–1;
    requirements of 242; results 120;
    theory 119–20
Coltrane, John 75
Columbia University 157
commensuration 2, 14–15, 22; acts of 22;
    protocols of 14; work of 2
Commentaries on the Motions of Mars
    (later titled *Astronomia Nova*) 198
compact muon solenoid detector *see*
    CMS
composite MPIs paired with graphs **94**
compressed sensing 73–85
compression 74, 76; algorithms 74; of
    data 20
computed tomography *see* CT
computer-aided tomography *see* CAT
concepts 2, 5, 25–32, 34–6, 106, 112–13,
    121–5, 148–9, 158, 210–12; abstract
    211; basic 30, 122; empirical 192;
    evaluative 106; many-faceted 35;
    mathematical 76, 78, 192; scientific
    27, 152, 204–5, 210–12; theoretical 34,
    211–12
conceptualization of measurement
    accuracy in blood pressure 138
confirmation, visual 112
confounding, variables 136, 138, 141–2,
    144
consistency 107, 123, 242
"constructivist implications" 118–19
content 3, 87, 91, 95, 115, 120–2, 135,
    155, 234–5; abstract 94; additional
    epistemic 211; empirical 28, 120–1,

123; neutral 120; observable 134, 138–9; pictorial 116; picture's 115; representational 88, 90; theoretical 208, 211; of visualizations 3
conventionalist approach, and the justification of rightness 117
Copernicus, Nicholas 189–92, 194–5, 198
coronal view 63, 65, 68
correspondence theory 117
correspondences 23, 32, 114, 117, 122, 190, 193; between the picture and the object depicted 117
cost functions 74, 79–81; external 74, 78, 85; nonfiscal 76
Courtenay, Nadine 216
Coville, Frederick Vernon 171–2
Craig, Patricia 177, 179
CRC Handbook of Chemistry and Physics 214–15
cross-disciplinary exchanges 171
CT 63–4
cultural assumptions 57, 63–4, 70

Dana, Harold W. 142
Darrigol, Olivier 216
Darwin, Charles 15, 19, 83
Daston, Lorraine 42, 44, 92
data and phenomena 206–8
data sciences 3, 81
Davidson, Donald 118, 120–4
Davis, Hallowell 75
Davis, Katharine Bement 157
Demenÿ, George 49
dendrograph 5, 170–1, 173, 175–81; additional 180; involving multiple 180; research 181; sales 180; studies 179; tested 178
DePace, Angela H. 101
depictions 2, 47–50, 53, 61–2, 64, 66–7, 87, 99, 117, 120; banned by society 47; and detection 99; distinct motive for 49; philosophical analysis of 87
Descartes, René 113
desert 170–1, 173–7, 179, 181–2; environment 171; harsh 175; heat 173, 175–6; real 175; remote 171
Desert Laboratory 170–82
design 43, 78–9, 81, 171, 179–80, 203–6, 209–10, 214, 216, 233–4; complicated 97; and creation 179; experimental 207; mechanical 195; multivariate 162; technological 206

diagrams 19–20, 91, 96, 111–12; conveying relatively abstract content 94; "eclipse" 191
Dialogues Concerning Two New Sciences 212
Dickinson, Robert Latou 157
direct measurement of transpiration 177
direct-reading instruments 236
Drucker, Donna J. 5, 157–66
Duhem, Pierre 233
Dumit, Joseph 57, 64–5, 67, 69
Dupré, John 30
Dutch National Science Foundation 216
dynamic change 174
dynamic integration 192
dynamical processes 192, 198

earth models 6, 220–8, 231; different 223; given 225–6; individuate 223; isotropic 228, 231; particular 224; unique 223
ecologists 174–6, 180–2; Carnegie-funded 178; community of 176; early American plant 174
effectiveness 144, 149, 151, 154; of psychometrics 149, 151, 153–4; and reality 154
Egger, Dave 79
Einstein, Albert 30–1
Emerson, Ralph Waldo 2, 11–15, 17–18, 21
empirical equivalence 224
engineering sciences 6, 203–16
epistemic practices 3, 100, 102, 104; scientist's 107; of testimony 107–8
epistemic sources 105, 108
epistemic things 18
epistemology 1, 204–5, 212, 216, 228, 249; contemporary 143; of economics 228; of measurement 1–3, 6, 82, 138, 233–9, 249; of producing knowledge of physical phenomena 203–4; of seismology 221, 228–9, 231; traditional 234
Eratosthenes of Cyrene 193
ethics 13, 20, 144
Eudoxus 193
evidence 3–4, 87–8, 91, 105, 135–6, 207–8, 220, 231, 234, 236; empirical 247; faked 124; historical 119; photographic 103, 124; visual 51; *see also* pictorial evidence
evidential role 4, 87, 101

experimental practices 209–10, 212–13, 216
experiments 59–61, 66–7, 70, 88–91, 93, 95, 175–6, 192, 208–12, 215–16; gel electrophoresis 89; giant ecological 174; independent 46; intensive 82; literary 18; new 209–10; paradigmatic 211–12; particular 89; second 60; theory-laden 82
explanatory use of visual data 100
exploration 6, 74, 85, 97, 162, 171, 228, 231; instruments for 231; models for 228; narratives 15; space 79
exploratory use of visual data 100
extension (of concepts) 36

Faulkner, Paul 105
Fechner, Gustav Theodor 152
Fisher, Martin H. 60
Fleck, Ludwik 101
Flynn, James R. 154–5
forests 18, 84, 175–6, 179–81
Fourier series 73–4
Fricker, Elizabeth 106–7

Gadamer, Hans-Georg 143
Galilei, Galileo 26, 49, 76, 193, 212
Galison, Peter 42, 44, 92
genes 89, 91, 93, 95; expression 88–91; known 91; monarch 91; noncircadian 91
Giere, Ronald N. 87
Gilbert, Freeman 225–6
Gingerich, Owen 193
Gleason, Henry A. 174
Glymour, Bruce 207–9
gold standards 133
Goodman, Nelson 87, 90, 111–12, 118–20, 123
Gotay, Carolyn 135–8, 142
graphic cartoons 64, 67
Guillaumin, Godfrey 5, 189–99

Hacking, Ian 112
Hamilton, Gilbert V. 157
harmonic analysis 73
Hausken, Liv 3, 57–70
Helmholtz, Hermann von 46
Hempel, Carl G. 27–8, 35
Henderson, Janice 191
Herbart, Johann Friedrich 147
Herschel, John 46
Herschel, William 46

Hipparchus of Nicaea 191–4
Hobart, Jeremy 133–5
Hölder, Otto 150
Holton, Gerald 26
homo mensura 41, 43
homosexuality 158–9, 162
Hopkins, Robert 116
Hubble, Edwin 48
Humboldt, Alexander von 15–16
Humboldtian science 16
Hunt, Robert 42
Hunt, Sonia 133
Hurter & Driffield's Actinograph slide-rule **45**
hypothesis-driven inquiries 87–91, 101, 194, 209–10, 216
hypothetico-deductive methods 223

idealizations 115, 224, 226–7, 236, 240; and abstractions 236; and approximation 240; processes of 115, 139
image interpretation 91
imaging technologies 3, 70, 87–8, 92–4
imagining 46–9
inferences 99, 115, 207, 214, 222, 230, 236, 242, 245; complex 236; statistical 207
information 3, 58, 60, 79, 93, 95–6, 100–2, 104–5, 220, 222; background 106; conceptual 57; continual 177; extracting of 61, 67, 226, 230; governmental 67; new 237; pertinent 242; public 65; relevant 4; stored 82; systems 44; technology 99; transmission 4; transmitting of 105
instrument indications 235–7, 239–41, 246
instrument manufacture 176
instrumentation 171, 174–5; ecological 170; technological 215
instruments 99–100, 102–4, 135, 176, 178–82, 192–8, 231, 236–7, 240–2, 244–8; adapting laboratory 181; for astronomical observations 196; calibration function maps 241; complex 175; custom-designed 176; dedicated 182; efficient 69; expensive 79; fine-tuning of 172; inventive field 171; mathematical 193; for measurement 231; musical 203; new 5, 53, 135, 171, 173, 176, 192; for observation 190; one-off 176; optical

117, 124, 199; for personal discovery
157; physical 228; preexisting
laboratory 176; scientific 170, 178–9;
specialized 181; technological 115,
214–15; traditional 198; *see also*
measuring instruments
integrative measurement 192
intelligence 4–5, 147–9, 151–5;
benchmark 155; high 153;
measurement of 5, 148, 154; research
147; research on 147; tests 151, 153–5;
theory of 149
intelligence quotient (IQ) 149
International Prototype of the
Kilogram, *see* IPK
interpretation 4, 42, 88–93, 97, 101–5,
113–15, 118–19, 134, 141, 212; and
human intervention 42; image
91; isolated 4; reciprocal 122; of
experimental findings 212; of scientific
images 118; scientist's 102; theoretical
215; of vertical spatial relations 89
inventions 5, 34, 46, 173, 178,
194; of Daguerre 42; important
photographic 46
inverse problems 222–3, 227–9, 231
IPK 244

James, William 33
Jeffreys, Harold 222
John, Mark St. 60
JPEG 3, 74–6, 82

Kaminski, Andreas 5, 147–55
Kant, Immanuel 147, 151
Keehner, Madeleine 57, 60–2, 65–7
Kemp, Martin 100–1, 105–7
Kepler, Johannes 5, 53, 189, 194–9
kilogram, (definition of) 244
Kinsey, Alfred 158–66
Kinsey Scale 165
Kjørup, Søren 116
Klein, Fritz 157
Klein Sexual Orientation Grid *see*
KSOG
The Klein Sexual Orientation Grid **158**
Knorr Cetina, Karin 88, 90, 92
knowledge 1, 5–6, 17–18, 82, 105–8, 111,
204–6, 208–10, 216, 233–4; accurate
196; antecedent 221; authoritative 57;
claims 233, 235–7, 246, 249; compara-
tive 178; conceptual 57; empirical 193,
229; generating of 210; lab pooling

90; objective 240–1, 248; optical 197;
reliable 148, 233; by testimony 105
KSOG 157–8, 161, 165
Kuhn, Thomas S. 118, 120, 125, 190

L1 minimization 76
large synoptic survey telescope 81
Latour, Bruno 14, 17, 21, 119
Lee, Maurice 22–3
length 14–15, 20, 25–9, 89, 141, 148,
151, 153, 213–14, 242–3; fragment 89;
measurement of 148, 151, 153; nucleic
acid 89–90; proliferation of concepts
of 27; standard 242; units of 214;
unknown 242
Lie, Merete 65, 67
light 2, 26, 42, 44, 53, 77–9, 90–1, 99,
195–7, 237–40; average 51; exposure
95; incandescent 78; microscopy 112,
119; natural 170; nature of 196–7;
period 91–2, 96; waves 26, 221
light in flight 52
Livingston, Burton 175–6, 178
Lopes, Dominic 116
Lord Kelvin 233
"lossy", and human visual systems 49
Lynn, Richard 154
Lynn, Sonya 164
Lynn-Flynn effect 149, 154

MacDougal, Daniel T. 171–2, 174–81
magnetic resonance imaging, *see* MRI
Mahler, Gustav 29
mapping functions 104, 107
Marey, Étienne-Jules **43**, 44–9
mathematization 204–5, 213–14, 216;
epistemic strategies of experimentation and 214; and measurements 204,
213, 216
Mayberry, Lisa 60
Maynard, Patrick 2–3, 41–53, 99, 103
McAllister, James 207
McCabe, David P. 59
McClimans, Leah 4, 133–44
McCloud, Scott 64, 66
meaning as use 28, 31
meaningfulness 2, 28–9, 31; of concepts
29, 31; loss of 29; of measurements 2;
well-rounded 31
measure of sexual identity exploration
and Commitment *see* MoSIEC
measurement 1–6, 25–32, 73–6, 78–82,
133–45, 147–55, 189–99, 203–16,

228–31, 233–44; accuracy 138, 140, 237–9, 243; accurate 171, 193; angular 194; compressed 76, 82; controlled 43; convenient 193; conventional 77; convergent 229–30; data 3, 115; densitometer 51; derived 151; devices 76, 78, 99, 102–3; errors 234, 238; faster 80; fundamental 151; initial 3; mediocre 193; mental 152; meridian 41; model-based 221; objective 43, 46; performing 73, 192; photographic 41, 70; photoluminescence 77; photometric 104; physical 246; planetary 198; random 75–8, 81; regularly-spaced 73; representational theory of 241; scientific 41–2, 151, 189, 192, 198–9; seismological 230–1; sexual 157; sharing design 46; single pixel 79; slit-scan 50; systematic 195
measurement outcomes 35, 214, 235–42, 245, 247–9; multiple inaccurate 249; putative 248; relevant 235; stable 233
measurement standards 244
measuring instruments 138, 141, 143, 192, 221, 228, 234–5, 239, 243–5, 248
mechanically produced images *see* MPIs
media 3, 57–8, 62, 68, 155, 230
media aesthetics 57–9, 61, 63, 65–7, 69–70
mental measure 150, 152
mental performance 147–8
mental properties 150–2
Merleau-Ponty, Maurice 62, 66, 68
meter, poetic 14
metric scale 42
metrology 234, 243
Michell, Joel 150, 152, 233
misinterpretations 112, 124
misrepresentations 112, 119, 125
Miyake, Teru 6, 220–31
model-based conception of calibration 234, 239
model space 225, 227
models 6, 84–5, 140–1, 193–4, 220–1, 223–31, 239–40, 242–3, 245–6, 248–9; accurate 242; black-box 243, 248; causal 85; and causation chains 84; complete astronomical 193; contemporary 158; general 57; idealized 229, 234, 240–1, 248; inaccurate 242; metrical 15; oversimplified 84; seismological 231; simplified 242; statistical

140, 244; theoretical 139; toy 84; universally-adjustable 181; untested 224; white-box 245, 248
MoSIEC 162
Mößner, Nicola 1–6, 99–108
movies: *Blow-Up* 82, 84; *The X Files* 84
MPEG 74, 76
MPIs 87, 91, 93, 95–7
MRI **63**, 64–5, 74, 79–81
Mt Wilson Observatory 173
mug shots 68–9
multiplicity (of measurement methods) 2
myth of absolute accuracy 243

NASA 49
neuroscience 3, 58–9, 62–3
non-nomological science 83
Nordmann, Alfred 1–6, 216
normativity, and the challenge of practice 237
North, John 191
Norvig, Peter 85

objectivity 3, 42–6, 48, 51, 92, 234, 236–7, 240–1, 248–9; in measurement 46, 48, 241, 248; mechanical 92–4
operationalism 2, 25–9, 31–3, 35–6; doctrine of 25; extreme 240; manifestation of 31; revival of 25, 32; significance of 35
operations 2, 26–30, 34, 78, 151, 248; empirical 151–2; experimental horticultural 176; instrumental 28, 31; mathematical 31, 151; measurement 26–9, 31–2; paper-and-pencil 28–9, 31, 35; physical 34
optics, analysis of 197
Origgi, Gloria 106–7
The Origin of Species 83

paper-and-pencil operations 28–9, 31, 35
Paralipomena ad Vitellionem. Astronomia pars optica 196
Pasteur, Louis 83–5
Pasteur's Quadrant: Basic Science and Technological Innovation 84
Peacocke, Christopher 116
Perini, Laura 3–4, 87, 87–97, 99, 103–5, 108, 112
persuasive mode 101
persuasiveness 57, 59
PET scans 63–5

Peterson's Quadrant 83–5
phenomena 2, 5, 21–2, 26, 29–30, 58,
 64, 69, 121, 204–16; behavioral
 59; cognitive 70; and data 208;
 empirical 13, 28; environmental
 177; interacting 206; mental 152;
 natural 43; physical 204–5, 208;
 purported 209, 212–13; unknown 204;
 unobservable 204–5, 208
philosophy of science 147–8, 208–9, 213,
 215–16, 234, 238
philosophy of technology 147, 234
photo finish 50, **51**, 52
photography 2–3, 41–50, 52–3, 67,
 99, 102–3; analysis of 2; composite
 68; computational 53; light in flight
 52; moving picture 47; nineteenths
 century portrait 69; for printing of
 scientific plates 42
photometric method 42, 102
physical phenomena 203–6, 208–10,
 212–13, 215–16; hypothesized
 210; purported 210; relevant 206;
 unobservable 204
physical properties 133, 140, 204,
 214–16; of materials 204; measured
 215; quantifiable 213
physiognomy 69–70
physiological ecology 175–8, 180–1
pictorial evidence 111–13, 115, 117, 119,
 121, 123–5; concept of 121, 124; a
 dynamic concept of 124–5
pictures 2, 4, 41–2, 49, 51, 64–5, 68, 70,
 102, 111–25; of brain activation 68;
 causal theory of 114; correctness of
 111–12; digital 124; invented motion
 47; manipulating 123; multi-colored
 46; naïve 204; perceived 117;
 photographic 103; scientific 111, 114,
 116, 121, 123; storing 80
plant physiology 171–2, 174–6, 182
pluralism 36
poetics 12–14, 17, 23
poetry 11–12, 14–15
Polanyi, Michael 66–7
portraits 42, 68–9; bourgeois 69;
 exhaustive ecological 174; of
 human beings 68; photographic 67–8;
 posed 45
positron emission tomography *see* PET
pragmatism 4, 33, 85, 114, 121–3
precision 23, 96, 190, 197; high
 instrumental 199; of instruments 197;

laboratory-style 182; measurement
 technology 181
predictions 33, 83–4, 195–6, 204,
 224, 242–3; deriving verifiable 216;
 inaccurate 247
Press, Frank 224
pressure 25–7, 141, 213; atmospheric
 212; diastolic 142; environmental 177;
 extreme 26; high 26
primary standards 138–40, 142
principle of charity 123
problems 104–6, 111–17, 120–1, 133–5,
 140–2, 144, 221–7, 230–1, 238–9,
 249; astronomical 193; classic 224;
 cognitive 196; communication 88;
 coordination 143; epistemological
 221, 225, 239; forward 222–3, 226;
 hermeneutic 144; inverse 222–3,
 227–9, 231; mathematical 225;
 methodological 84; nonuniqueness
 223, 226, 228; practical 148; primary
 14; recurrent 160; salient 58; of
 scientists 248
properties 33, 83, 102, 113–16, 151–2,
 204, 210–15, 220–2, 226–8, 235;
 emergent 21; interior 222; local
 246; material 211, 215; mechanical
 221–2, 224; new 215; and phenomena
 213, 215; physical 133, 140, 204,
 214–16; real 112, 115, 117; relevant
 atomic 246; unified 27; well-defined
 mathematical 27, 73
Protagoras 41, 43–4
protocols 13–14, 16, 18, 21–2; basic 17;
 of commensuration 14; contriving
 2; of measurement 22; new 14; of
 Thoreau 17
psychology 60, 147–9
psychometrics 5, 147–55
psychotechnics 149
Ptolemy, Claudius 5, 189–94, 198
public perceptions, of brain imaging 67
publications: *Bisexual Option* 157, 159,
 166; *Commentaries on the Motions of
 Mars* (later titled *Astronomia Nova*)
 198; *CRC Handbook of Chemistry
 and Physics* 214–15; *Dialogues
 Concerning Two New Sciences* 212;
 Journal of Homosexuality 162; *The
 Origin of Species* 83; *Paralipomena
 ad Vitellionem. Astronomia pars
 optica* 196; *Pasteur's Quadrant: Basic
 Science and Technological Innovation*

84; *Seeing New Worlds* 16; *Sexual Behavior in the Human Male* 159; *A Week on the Concord and Merrimack Rivers* 11, 22

quality of life 4, 133–44; instruments 134, 136–7, 140, 142; measurability of 4, 143; measurement 134, 142–3; and measurement challenges 136; measurement interaction 144; measures 133–5, 138, 140, 144; research 133, 136, 140, 142; researchers 5, 133–5, 137, 139, 142; and response shift 143–4
questionnaires 1, 5, 134, 137, 139, 143–4, 166

radiation exposure 76, 80
radiometry 44
RC 140–1
reality 2, 33–5, 66, 118, 121, 144, 154, 175, 240; definition of 33; direct mental apprehension of 34; distorted perception of 144; external 33; notion of 33; physical 34, 197
realizations 76, 140–1, 244; metrologically-sanctioned 246; official 245; particular 244; primary standard's 140
recording 1–3, 12, 43, 45–6, 53, 177
reference, pictorial 112–16
regulative principles 205, 209, 214
relational ontology 21
relativism 43, 120, 125
representation 90; dependent 116; practice 119; and rightness 123; of space 62
reproduction 46–7, 65, 67, 87, 114, 119, 121, 123
reputation 4, 106–7, 155, 178 enhanced scientific 5; individual 107; scientist's 106, 108; and social roles 107; testifier's 107
resemblance 47, 60, 62, 104, 107, 113–14, 116–17, 122; external 113, 116–17; internal 113, 116–17; pre-existing visual 47; representation-dependent 116, 123; representation-independent 116–17, 123; theory 116; visual 47, 60, 62
response shift 136–40, 142–4; conceptualize 137–8, 144; obscure 138
Rheinberger, Hans-Jörg 18, 20, 120–1

Richter, Charles 222
rightness 4, 111–14, 117–19, 121, 123–5; conditions of 121; dynamic 114; objective 113–14, 116–17, 119; of pictures 4, 111, 124; representational 123; subjective 119, 121; tentative 125
robust perspectivism 248–9
robustness condition, *see* RC
romantic literature 11, 152
Rorty, Richard 147
Roskies, Adina L. 57, 67–8
Russell, Bertrand 119

sagittal view 63, 65, 68
San Diego 157
San Francisco 164
Schaffer, Simon 21
scheme 83, 118, 120–1, 125; classification 83; conceptual 120; known transform coding 75; relativity 120–1; unique 117–18
Schöttler, Tobias 4, 111–25
Science 178
scientific concepts 27, 152, 204–5, 210–12
scientific knowledge 83, 91, 189, 199, 205, 209, 216; application of 205; generating 87; of physical phenomena 205, 208–10, 212, 214, 216; production of 214
scientific measurement 41–2, 151, 189, 192, 198–9
Seeing New Worlds 16
seismology 220–1, 223, 225, 227–31
self-registration 42–3, 45
sensitometry 44
sexual behavior 158–9, 166
Sexual Behavior in the Human Male 159
Sexual Behavior in the Human Female
sexual identity 159–64, 166; measure 162; personal 160–1; political 161
sexual identity tests 163
sexuality 157–8, 161–3, 166; conceptualize 160; and emotional/social preference 162; person's 162; unlink from identity 165
Shannon-Nyquist theorem 74, 76, 81–2, 85
similarity 87, 114, 116–17, 122–3; formal 68; representation-dependent 117; visual 62
single pixel camera 12, 77, **78**, 79, 81
slit-scan 50, 52

Sloan Digital Sky Survey 81
Smallman, Harvey S. 60
software epistemology 82
sparsity 73–7, 82
statistical analysis 159, 207
Stern, William 149
Stevens, Stanley S. 1, 75
Stokes's Quadrants **84**
subjectivism 44, 117
subjectivity 42, 51, 114, 149, 153–4;
    conceptualize 152; potential 153;
    sexual 161
surrogates 3, 100–2, 111, 116
surveying, professional 16, 22
Swerdlow, Noel M. 190
Sykes, Godfrey 170–4, 176–82
symbols 4, 18, 95, 114, 117–23, 235

Tal, Eran 6, 138, 233–49
Talbot, William Henry 42, 46–7
Taylor, Charles 14
technicians 5, 171–2, 176
technologies 2–3, 41, 46, 53, 65, 67, 99,
    103, 147–9, 204–6
technologies of the senses 58
testimony 105, 107–8, 111
Thagard, Paul 120
theory change 189, 198
theory-ladenness 82, 208
thermometers 11, 16, 42, 152–3, 233,
    236, 240; calibrated alcohol 241;
    divergent 248; household mercury
    236; Jordan's 52; readings 206;
    single 206
third dogma 120
Thoreau, Henry David 2, 11–23
time 26, 42–4, 46, 49–53, 136–8, 164,
    173, 190–1, 193–4, 220–2; absolute
    31; ancient 191, 197; clocked
    exposure 42; data collection 79;
    new standard 22
Torpey, John 69
transaxial view 63, 66, 68
Trendler, Günter 150
triangulation 121–5
"True Meridian" 19

"True North" 20
Tucson 170, 172–3, 176–7, 179,
    181–2

Udis-Kessler, Amanda 160, 166
uncertainty 140, 223, 226, 247; amount
    of 139; margin of 235; measurement
    of 234, 243; and modeling in
    seismology 220–1, 223, 225, 227,
    229, 231
underdetermination 76, 224, 247
underdetermined linear systems of
    observations 76
understanding 43, 64, 111, 124, 243;
    dominant cultural 166; of ecological
    relationships 174; historical 193; of
    pictorial evidence 121; of Thoreau's
    protocols of measurement 17
unity 35–6
USA Track & Field photo finish **51**

validity 32, 52, 104, 135–6, 141, 144,
    150, 162
van Fraassen, Bas C. 32, 87, 136, 143–4,
    212
van Helden, Albert 191, 193, 195
visualizations 3, 62–5, 99–105, 113–14,
    123–4; of brain imaging data 64–5;
    epistemic status of 102; scientific 3, 57,
    61, 63, 100, 102, 116; software 2; of
    staining cells 63; techniques 63–4
Vogt, Thomas 3, 73–85
Vögtli, Alexander 101

Walden 11–12, 15, 18, 20–3
Walden Pond **12**, 13–14, 19–20
Walls, Laura Dassow 1, 11–23, 175
wavelets 73–7
Weber, Ernst Heinrich 147
*A Week on the Concord and Merrimack
    Rivers* 11, 22
Weisberg, Deena Skolnick 57–9, 68, 70
Wellcome Image Awards 63
Wilder, Kelley 103
Wittgenstein, Ludwig 28, 33, 124
Woodward, James 204, 206–9, 213

Printed and bound by CPI Group (UK) Ltd, Croydon, CR0 4YY

21/10/2024

01777087-0016